Environmental Degradation and Global Health

The Editor

Dr. Ashwani Kumar Dubey is President of Environment and Social Welfare Society, Khajuraho and Editor-in-Chief of "International Journal of Global Science Research, India". He is currently working as Guest Lecturer in Department of Zoology, Government Maharaja Post Graduate College, Chhatarpur, Madhya Pradesh.

He did M. Sc. in Zoology with specialization of Ichthyology in 1991 from Awadhesh Pratap Singh University, Rewa Madhya Pradesh. He obtained his Ph.D. Degree in 1995 entitled "Responses of antioxidants, lipid-protein interactions and lipid peroxidation in *Heteropneustes fossilis* to oxidative damage exposure" from School of Studies in Zoology, Vikram University, Ujjain, Madhya Pradesh, India. His research field was Biochemistry, Free Radical Biology, Toxicology and Stress Monitoring. He has published several research papers in Foreign Journals including Pergamon Press United Kingdom to his credit. He published many popular science articles, delivered lectures and broadcasted science talk by all India radio for the students and social welfare.

He joined as a Scientist, Research and Development in the Rank Industries Ltd., Nellore, in Andhra Pradesh State in 1995. During this service period he served for Pathological and Water Quality Assurance in Aquaculture Division. Then in 1998 he was appointed as an Assistant Professor of Zoology at RBS College, Rajnagar. He managed the College through a major transformation of its research and teaching. He played a major role in the design and construction of a new Fisheries Demonstration Centre at Godavari Estate, Nahdora Khajuraho, India. In 2004 he became Guest Lecturer of Zoology, in Government College, Maharajpur, Higher Education Madhya Pradesh. For most of his career in research interests have focused on the Biodiversity, Environmental Impact Assessment and Bio-Resources Conservation.

Under the able guidance of executive director Dr. Ashwani Kumar Dubey, the ESW Society has carved the niche through various activities, programs and awareness camp to promote the cause of environment and social welfare. Apart from this, the Society established Godavari Academy of Science and Technology in 2010 to promote scientific awareness for environmental conservation awareness through seminar, workshop, conference and symposium. To support the cause of environment awareness through scientific thinking and reasoning, the society has started the international journal of global science research (IJGSR) providing a huge platform for academic interaction through research papers and analytical interpretation.

Dr. Ashwani awarded by many prestigious awards by National and International institution. He is a regular member and fellow of important scientific societies and also in editorial board member of Research Journals in India, America, United Kingdom, France, and Japan

Currently: Awarded by **National Excellence Award-2015** for outstanding contributions in education for socio-economic development for the Nation by ALMA London, United Kingdom International Star Performance Certification 22 February, 2015.

Dr. Ashwani is currently working on Impact of environmental degradation, global health, natural resource, wildlife conservation and stress monitoring.

His personal interest includes Reading, Writing, Traveling and Photography.

Environmental Degradation and Global Health

Editor

Dr. Ashwani Kumar Dubey

2016

Daya Publishing House®

A Division of

Astral International Pvt. Ltd.

New Delhi – 110 002

Cataloging in Publication Data--DK

Courtesy: D.K. Agencies (P) Ltd. <docinfo@dkagencies.com>

Annual National Conference on "Environmental Degradation and Global Health" (2nd : 2015 : Khajuraho, India)

Environmental degradation and global health / editor, Dr. Ashwani Kumar Dubey.

pages cm

One chapter in Hindi.

Papers presented at the Annual National Conference on "Environmental Degradation and Global Health" organized by Environment and Social Welfare Society, held at Khajurāho, India during 1-2 February 2015.

Includes bibliographical references and index.

ISBN 978-93-5130-881-2 (International Edition)

1. Environmental degradation–India–Congresses. 2. Public health–India–Congresses. I. Dubey, Ashwani Kumar, editor. II. Environment and Social Welfare Society (Khajurāho, India) organizer. III. Title.

GE160.I4A56 2015 DDC 363.700954 23

Published by : **Daya Publishing House®**
A Division of
Astral International Pvt. Ltd.
– ISO 9001:2008 Certified Company –
4760-61/23, Ansari Road, Darya Ganj
New Delhi-110 002
Ph. 011-43549197, 23278134
E-mail: info@astralint.com
Website: www.astralint.com

Laser Typesetting : **Classic Computer Services,** Delhi - 110 035

Acknowledgement

The National conference on *"Environmental Degradation and Global Health"* has its inception in 2013 when a disaster in India as well later in Japan as in other parts of world was observed due to a natural calamity "Tsunami" brought a vast change in minutes as the thrusty area for work in Environmental Degradation and Global Health provide a field for research and discussion.

It is my privilege and pleasure to express my profound gratitude to our Chief Guest Prof. Vinod Prakash Sharma, NASI-ICMR Chair Distinguished Professor (Public Health Res.) Centre of Rural Development and Technology, Indian Institute of Technology Hauz Khas, New Delhi. Guest of honour Dr. Sneh Bhargava, Professor Emeritus and Formerly Director, All India Institute of Medical Science, New Delhi. Dr. S. N. Pandey, Vice Chancellor, Indira Gandhi Technological and Medical Sciences University, Ziro, Arunachal Pradesh, India, Prof. Kaidar Nath Singh Yadav, Vice Chancellor, Rani Durgavati University, Jabalpur, Madhya Pradesh who have given very kindly consented for Inaugural Programme of National Conference.

I am Thankful to Bundelkhand Extended Region Chapter, Chitrakoot, The National Academy of Sciences India, Allahabad for its in association with Environment and Social Welfare Society, Khajuraho for organizing this Conference.

I am highly thankful to Prof. Pramod Kumar Verma, Scientific Advisor, Government of Madhya Pradesh and Director General, Madhya Pradesh Council of Science and Technology, Bhopal for grant to Environment and Social Welfare Society, Khajuraho for this National Conference.

I am thankful to Honorable Guest Prof. U. C. Srivastava, Department of Zoology, University of Allahabad, Allahabad UP. Dr. Niraj Kumar, Executive Secretary, The National Academy of Sciences India, Allahabad. Dr. N. P. Shukla, Chairman, Madhya Pradesh Pollution Control Board, Bhopal. Dr S. K. Mandal, Chief Conservator of Forest, Range Chhatarpur, Dr. Srinivas Murthi, Chief Conservator of Forest, Panna

Tiger Reserve, Panna. Dr. Masood Akhtar, Collector and District Magistrate, Chhatarpur. Dr. Rajesh Saxena, Sr. Scientist, Madhya Pradesh Council of Science and Technology, Bhopal M.P. and Mr. Susheel Varmon, District Co-ordinator, JAP, Chhatarpur for his valuable support and inspiration.

I am heartily thankful to honorable Invitee Guest Who have very kindly consented and given us an opportunity to share valuable thought which will provide milestone on the way of leading Scientists in this National Conference.

I am heartily thankful to Dr. R. S. Chauhan, Formar Director, Deptt. of Aquaculture, College of Fisheries, G B Pant University of Agri. and Tech., Pantnagar, Uttarakhand who have very kindly consented for Valedictory Function and Award Ceremony.

I am especially thankful to all delegates who actively participated in this National Conference.

I am profoundly thankful to my Board of Director and All members of Environment and Social Welfare Society, Khajuraho for their invaluable cooperation, and those entire person who are directly or indirectly concerned with this conference.

Dr. Ashwani Kumar Dubey

Chief Guest's Address

I am greatly privileged to be part of this national conference on "Environmental Degradation and Global Health" and thank the organizers for giving me an opportunity to be its part.

The industrial revolution of the nineteenth century, no doubt, has considerably improved our living standards, health care and availability of novel materials and processes. These improvements, however, have triggered a massive consumption of energy (fossil fuels) and natural resources resulting in a significant degradation of natural environment and greater emission and accumulation of greenhouse gases (for example, carbon dioxide, methane, CFCs, water vapors etc.) in the biosphere. These greenhouse gases are responsible for the unique problem of global warming threatening the very survival of the human society and its resources. Global warming does not affect only the climate and extreme weather events but will also facilitate shift in biotic diversity and activity from lower altitudes and latitudes to the higher ones, which may result in social tensions, widespread loss of biodiversity and natural resources and disturbance in the natural balance of the Earth.

India is the world's fourth largest economy and fifth largest greenhouse gas (GHG) emitter, accounting for about 5 per cent of global emissions. The United Nations Convention on Climate Change and its Kyoto Protocol have strived to address this unique problem through forging mutual agreement and understanding in lowering the emission of greenhouse gases and adoption of new environmental friendly technologies for the last 30 year with limited success. The fifth Assessment Report (AR5) released in 2014 and the Synthesis Report released in 2014 by the Intergovernmental Panel on Climate change has established clearly the impending dangers of the global warming for the human society with hard facts and data. According to these reports, the global average air temperature near the Earth's surface has already risen by 0.74 ± 0.18^0C during the hundred years ending in 2005 and may rise a further 1.1 to 6.4°C by the end of this century. According to latest reports, 2014 was the hottest year since 1880. The average temperature of the year was 14.6°C, 1.24°C above the 20th century mean.

The climate changes are expected to happen gradually, but are already considered responsible for occurrence of extreme weather events such as heavy downpours and flooding, heat strokes and waves, intense hurricanes and storms, and drought and famine. The latter events have dire indirect and direct impacts on the human health. The adverse health effects associated with climate change will likely be exacerbated in low and middle-income countries. High risk areas include the ones already experiencing scarcity of resources, High incidence of infectious and vector borne diseases. In India, Increased frequency of droughts, floods due to greater intensity of rainfall events glacier lake outburst floods (GLOFs) and tropical cyclones will further aggravate human health due to breakdown in public health infrastructure and services. Some prime examples already provided the image of the future like.

1. European Heat Wave (2003) killing thousands of people.
2. Hurricane Katrina (2005) killing about 1800 people and displacing thousands in the united states besides paralyzing the existing public health and sanitary infrastructure.
3. Cyclone Aila (2009) killing about 700 people and displacing millions in Bangladesh, India and Myanmar.
4. Cyclone Phailin (October 2013) killing only 1 person and causing losses to the tune of Rs. 3000 crore near Goopalpur in Orissa.
5. Cyclone Hudhud (12 October 2014) killing about 109 people and causing extensive losses to the infrastructure and property tune of Rs. 70000 crore near Vishakhapatnam in Andhra Pradesh.
6. The 2013 GLOF in the Kedarnath region of Uttarakhand state in India killing thousands of pilgrims and other people.
7. Noticeable increase in the incidence of vector-borne (Japanese encephalitis, malaria, dengue and chikunguniya), water-borne (Cholera, diarrheas), air-borne (asthma, pneumonia) and parasitic (leishmaniasis, lymphatic filariasis and onchocerciasis) infections in the developing worlds of Africa and Asia.

The sizeable portion of human societies are already exposed to the climate-sensitive diseases with noteworthy examples of

1. Malnutrition causes in excess of 3.5 million deaths every year throughout the world.
2. Diarrheas kill more than 1.8 million people every year in the world.
3. Malaria and other vector-borne diseases are responsible about 1.2 million deaths per annum in the world.

For offsetting these environmental health hazards and risks, there is an urgent need to adopt the preventive methods by strengthening public health and sanitary infrastructure both in developed and developing nations. The events like cyclones, floods, hurricanes, storms etc. not only cause higher intensity of human mortality directly but also exacerbate the collateral damages by totally breaking down the existing public health and sanitary infrastructure. The ongoing inter-governmental negotiations to reduce emission of greenhouse gases may still take a long time to reach to a mutually agreed formula. In the meantime, it should be a better idea that the public be sensitized about the health hazards and prevention methodologies through well-designed awareness programs and the administrators be trained for effective management of the climate-sensitive diseases during extreme weather events.

I am optimistic that this two-day national conference will be able to explore new ideas, techniques and methodologies to understand the mechanisms behind the above-mentioned problems in future to come. I further hope that outcomes incurred from discussion by the learned scientists/ social scientists gathered here would be able to solve this complex problem prevalent in the society. I wish a grand success of this conference.

Prof. K.N.S. Yadava,

Vice-Chancellor,

Rani Durgavati University,

Jabalpur, Madhya Pradesh

Editorial

This is an honor for the **Environment and Social Welfare Society,** Khajuraho, Madhya Pradesh to organize the 2nd Annual National conference on **"Environmental Degradation and Global Health"** in the world heritage, Khajuraho, In Association with Bundelkhand Extended Region, Chapter, Chitrakoot, the National Academy of Sciences India, Allahabad. It is sponsored by Madhya Pradesh Council of Science and Technology, Bhopal. Assisted by Godavari Academy of Science and Technology, Chhatarpur.

The principal goal of this conference will be to present some of the latest outstanding breakthroughs in Environment and global health, to bring together both young and experienced scientists from all regions of the world, and to open up avenues for research collaborations at regional and global level. With theme taking some positive steps towards improving our Earth for future generation.

Objective of this conference is to provide a platform to Educational Administrators, Vice-chancellors, College Principals, Deans, Readers, Head of Departments, Professors, Assistant-Professors, Scientists, Environmentalist, Researchers, Young scientists and Students to disseminate knowledge related to Impact of "Environmental Degradation on Global health" the general topics covered in the conference will be as under:

Global Health Scenario

Natural calamite, Volcano, Natural disaster, Water conservation, Land degradation and Forest Conservation, Natural Resources and their Conservation, Green energy, Global warming, Social implication caused by alteration of bio-resources, Hydro-chemical and water quality, Recycling process of pollutant, Agrochemical and environmental hazards.

Environment Scenario

Oxidative Stress, Strategic Environmental Approach and Sustainability, Tribal environment relationship and community development, Environmental impact assessment, Preventive measures of environmental degradation, Environmental toxicity, Environmental resources and management, Environment Conservation and Validation of traditional knowledge, Pollution.

Wildlife and Biodiversity Scenario

Wild life Management System, Climate change and endangered species in India, Biodiversity and conservation, Conservation of critical and fragile habitats and corridors, Conservation and promotion of Medicinal plants, Sustainable Tourism, Ethno-biology and human welfare, Role of N.G.O. in Nature conservation

Technological Approach

Application of bio-technology, Rural bio-technology, Tools and technique: for protection and conservation of bio-resources, Bio-markers with special reference to climate change, Ecosystem and conservation measure

Environmental degradation is a result of socio-economical, technological and institutional activities. Degradation occurs when The Earth's natural resources are depleted. Air, water and soil resources are affected. Degradation also impact on microorganism, plant, animal and wildlife.

Our land is compromised when people exhaust resources or release harmful chemicals into the air. Deforestation, wasting resources, and pollution all add to the demise of an environmentally-sound and safe planet. Trees of forest are cut down in large quantities, so that more homes can be built on the land, the birds and wildlife who lived in the forest must find a new place to live. The vegetation that once grew on the land is destroyed. Wildlife is the essential part of the environment. The existence of wildlife is the most important aspect to maintain the balance between life and environment. Wild animals are facing threat of extinction due to environmental degradation.

The degradable changes observed in the environment, natural resources and biodiversity can be checked and renovated if each one of us take some sensitive and positive measures as the solution to save the earth from degradation caused by global warming, climatic changes, pollution, acid rains and natural calamities as well, therefore it is urgent to discuss and to find a possible solution related to environmental degradation.

Dr. Ashwani Kumar Dubey

Contents

	Acknowledgement	*v*
	Chief Guest's Address	*vii*
	Editorial	*xi*
1.	**Environmental Degradation and Global Health** *Kavita Chaudhary and R.K. Singh*	**1**
2.	**Role of Cooperatives in Environmental Protection** *Bibhu Santosh Behera, Rudra Ashish Behera, Anama Charan Behera and Gurudev Sahu*	**19**
3.	**Temperature Rise in Jabalpur in Last Thirty Years** *Aruna Pande and Rajesh Wahane*	**25**
4.	**Role of NGOs in Nature Conservation** *Dilip N. Pandit*	**33**
5.	**Enhancing Community Based Food and Water Security through Micro-Diversion Based Irrigation Initiative in Poverty Stricken Areas of Orissa** *Bibhu Santosh Behera, Rudra Ashish Behera and Anama Charan Behera*	**41**

6. **Repercussions of Mobile Phone Addiction among Teenagers: Special Study among Students of Selected Schools of Indore City of Madhya Pradesh** 51
Neeta Deshmukh

7. **Stored Grain Insect Pest and their Effective Ecofriendly Management** 59
Bibhu Santosh Behera, Soubhagya Behera and Subhashree Dash

8. **Determination of Water Quality of Surface Water in Eight Blocks of Banda Region: Chemical and Biological Studies** 65
Chhavi Purwar and Preety Agarwal

9. **Assessment of Ganges River Water Quality at Rani Ghat, Kanpur** 77
Padma Saxena

10. **Plankton Diversity Studies of Upper Lake with Reference to Physico-chemical Status of Water** 83
Sonam Gupta

11. **Quality Assessment of Groundwater in some Rural Areas of Kanpur** 93
Atul Kumar Misra

12. **Vulture Identification, Conservation and Awareness in Tikamgarh District, Madhya Pradesh** 101
Jagdish Prasad Rawat

13. **Impact of Environmental Degradation on Fisheries** 109
Ashwani Kumar Dubey

14. **Biotechnology and IPR** 113
Bharat Kumar Singh and Shreaa Nath Singh

15. **जलवायु परिवर्तन एवं ग्लोबल वार्मिंग** 121
मयूर शिखा अग्रवाल

Index 125

Chapter 1

Environmental Degradation and Global Health

Kavita Chaudhary[1] and R.K. Singh[2]

[1]Department of Microbiology, Govt. P.G. College, Noida, U.P.
[2]Department of Chemistry, Pt. J.N.P.G. College, Banda, U.P.

ABSTRACT

In the present scientific scenario there has been enormous improvement in living of humans but due to rapidly increase in the industrialization and materialistic culture, our environment is very badly polluted. It is true that we live in it, as such are bound to face hazard. According to IPCC report, climate change has taken a broad alarming situation and now we can see its effects around us. One of its serious cause is global warming. If we do not take steps to improve the environment, there is no doubt that nature will not gives us another chance. At present the earth's temperature is 1.2°C, but if this situation continues than after this century, we will have to face temperature increase upto 4°C. Such increase in the temperature will be equal to a new world war. Due to this rapid increase in the world's temperature, humans have to face a havoc type of situation in which there will be disastrous flood and draughts just like happened in Jaisalmer and Cherrapunji (Assam) respectively. Due to this temperature rise the glaciers in Himalaya region will start melting which will lead to an increase in the water levels of oceans, people have to face Tsunami's and Elninos. If this situation continues than, that day is not very far when them forever Ganges will stop its flow and become quite. The main reason for this cause would be the harmful green house gases like CO_2, N_2O, CH_4, CFC etc., which are increasing the earth's temperature. At present in developed countries the CO_2 release per man per year is 11 Metric ton, where as in developing countries it is 3 Metric ton. The large scale cutting of forests is also increasing the temperature of the earth day by day. Reducing area of forests now become one of the major factor of flood and drought. Thus like other pollutants, cutting of forests, increasing no. of vehicles, industries now become the important cause of high amount of CO_2 and SO_2 in the atmosphere. To control it the UNO is taking account as seen by the Kyoto Protocol declaration, 1977 for environment protection, which is very important. According to this taking standard of 1990, the total decrease of

GHG release is fixed upto 5.2 per cent. With respect to this, we can also change our life style by not adopting materialistic culture and can contribute to this movement. We must focus to take serious measures which include making of policies to reduce the release of GHG. Instead of using private vehicle, we should encourage use of public transport, We should also encourage the use of wind and Solar energy.

Keywords: *I.P.C.C.: Intergovernmental panel on climate change, G.H.G.: Green house gases; DDT: Dichlorodiphenyltrichloroethane; U.N.: United Nations; CO_2: Carbon dioxide; Mt.: Metric ton; N_2O: Nitrous oxide; SO_2: Sulphur dioxide; CFC: Chlorofluorocarbons; O_3: Ozone; N: Nitrogen; CO: Carbon monooxide; Hg: Mercury; H_2S: Hydrogen sulphide; Pb: Lead; Mn: Manganese; EPI: Environmental Pollution Index; K: Potassium; UV: Ultra Violet; ODS: Ozone depleting substances; CFL: Compact Fluorescent Lamp; pH: Parameter of Hydrogen.*

Introduction

First of all, let us understand what environmental degradation is and what its impact would be on the health of human beings globally. Undoubtedly we have prospered materially in the present age of science and technology. However, industrialization has also brought in pollution of every kind, *i.e.* air pollution, water pollution, sound pollution and nuclear radiation has increased unimaginably. IPCC a U.N. body to keep watch over climate change, has given a detailed report containing 2610 pages. It was prepared by 309 eminent scientists and has been released recently. It is clear premonition to all of us. The report clearly indicates that the climate change has taken minatory shape and its consequences are apparent. And if now we do not take the notice of such threaten, we are not going to get another chance to survive. It is very alarming that the developed countries are not taking Kyoto protocol 1997 seriously. According to this pact the target of GHG release should have to reduce up to 5.2 per cent. The name green house was given by **Arrhenius** in 1890. Here, it is necessary to mention that increase in the temperature globally in due to G.H.G. release and if there is no G.H.G. release. The temperature may go down up to -18°C. But excess rise in temperature is very harmful for global health. Here it will not be exaggerating to mention that we celebrate 5th June every year as world's Environmental day and 16th September as Ozone day.

Materials, Methods and Discussion

Today air pollution has become a serious problem as due to this about 30 million people die every year. During three decades the CO_2 release in atmosphere has been doubled. Developed countries of the world contribute about 10 Mt. of CO_2 in Composition to only about 3 Mt. by the developing countries. India represents the third world's greatest economy after U.S.A. and China. At present the world's population is about 6.25 billion which is likely to become a 9 billion in 1950 (Table 1.1). India's population is about 20 per cent (1/5) of the world's total population and it emits about 5 per cent of world's total release of CO_2, N_2O, SO_2, CFC etc. Where as USA and China's contribution is 16 per cent. At present the release of various gases, which are responsible to increase the temperature of earth are as follows : CO_2 = 49 per cent, CH_4 = 18 per cent, CFC = 14 per cent, N_2O = 6 per cent, Halogens, SO_2, O_3 =

13 per cent (Tables 1.2 and 1.3). Thus we see India stands 3rd after U.S.A. and China as for as the G.H.G. release is concerned. The entire population of the world is facing the shortage of clean drinking water. Area of Tropical, evergreen and rain forests is gradually decreasing and about half of the wet and marshy area of the world have vanished. About 1/5 of total 10,000 drinking water reserves are almost exhausted. We know that upper 9" layer of earth is very useful for cultivation and is fertile. Thousands of years are needed for the development of its quality. Due to heavy rain the upper layer of the fertile earth is washed away as such 2/3 of the fertile land has been destroyed. It is so precious when it is said as "the water of the field is actually the field itself." When man thought for better yield of crop he decided to use more and more chemical fertilizer which soon decreased the fertility of the soil. (an ideal soil has pH from 6.5 to 7.5). When the natural dung is quite useful and fit as fertilizer, why we use chemical fertilizer. This is matter of concern that why we use artificial fertilizer as how it is necessary to use chemical fertilizer. Although some experts suggest to mix small amount of urea in the dung, which is expected to increase the N content in the soil. But now every farmer have started using only urea and forgotten the use of dung.

Table 1.1: Increase in Population of World and India (In Billion)

	World		*India*	
Year	*2014*	*2050*	*2014*	*2050*
Population	6.25	9	1.25	1.8
Land Area (1000 Square K.M.)	135800	135800	3287	3287

Table 1.2: Emission of C.F.C. Gases

Country	*Amount of C.F.C. in Billion Mt. of Carbon*
U.S.A.	350
Japan	100
China	32
India	0.7

Table 1.3: Action Increases Earth Temperature (In per cent)

Energy uses and production	57
C.F.C.	17
Forming	14
Change of land use	9
Industrial use	3

About more than half of the meadows of the world have been finished and at least 80 species of plants have vanished in India and about 1000 are more to vanish.

We are looking a clear threat to biodiversity around us as every year about 11 million hectares forests are being vanished from global map and its adverse effect in seen in flood, drought, land slides, earth quake like natural calamities around us. Due to urbanisation, two out of three species are vanished. In India among 45,000 plant species and 75,000 animal, 79 species mammals, 44 of birds, 15 of reptilians, 4 of bipartite are about to vanish. Due to blind race of industrialization and development, the entire world is facing the following problem today.

Global Warming

When the temperature of the earth is more than the normal temperature, the phenomenon is called global warming. The changes occurred due to this effect are known as climate changes. Thus global warming in the consequence of climate change. The present temperature, 1.2°C of the earth is expected to rise up to 4°C after the end of 21th Centuary. Such situation is so dangerous like an world war. Due to which problems such as excess raining or drought would happen. As seen in the present scenario we encounter heavy snow fall as well as drought due to temperature rise. Two year earlier, number of people died in Russia due to heat stroke, which is known as a very cold state. Similarly in Pakistan thousands of people died to flash flood recently (Figure 1.1). Huge loss of lives has been seen in Thailand and Bangladesh due to heavy rain and flood. Flood in U.K. where as drought in U.S.A. is recently seen (Figure 1.2). Britain has encounter the dust storm from Sahara desert (Figure 1.3). Cherrapunji (Figure 1.4) which is known as the area of highest rain fall has to drought where as the desert like Jaisalmer (Figures 1.5–1.8) is seen dipped in flood. Moreover it has to face the problem of water drainage.

Figure 1.1

Figure 1.2

Figure 1.3

Due to increase in temperature the cities which lies on shore *i.e.* Mumbai, Dhaka, Calcutta are having a danger to sink as the sea water level is increasing. Due to this the countries like Maldives are planning to shift some where else. According to the

Figure 1.4

Figure 1.5

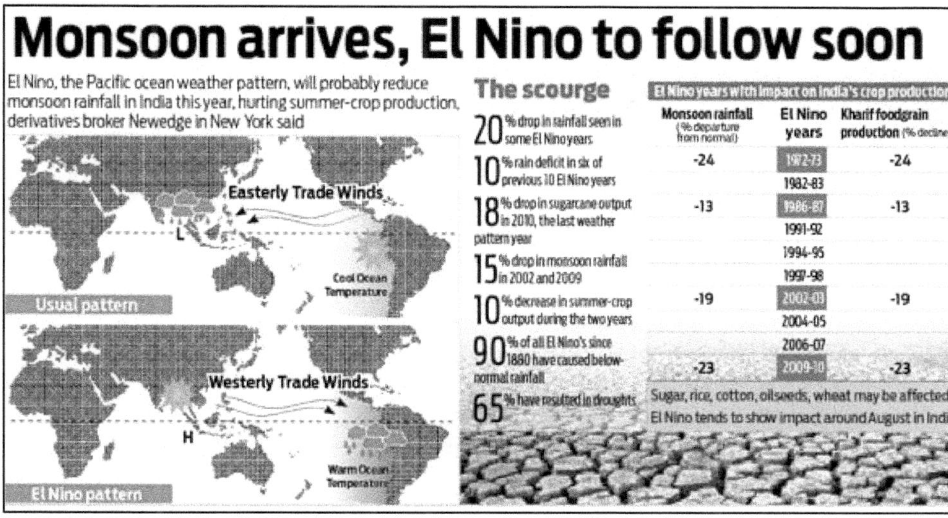

Figure 1.6

Figure 1.7

opinion or estimation of International organization for Migration until 2050 about 20 crore people must have migrated to safer place because of this problem (Figures 1.9 and 1.10). We can also see a decrease in monsoon rain and on the other hand in March and April we can see heavy rain and hailstone frequency Katrina, Elnino and

Figure 1.8

Figure 1.9

Tsunami Typhoon can also be seen very continuously. The heating of top level sea water on pacific ocean is known as Elnino (Figure 1.6) one to which the monsoon winds which are coming to India change their route to pacific ocean. So it effects the rain in this region. In this way changes in world's temperature effect the developed and developed countries. But it influences the developing countries more become these countries due to shortage of resources are bound to face such type of nature excess raining or short of raining due to global warming. As such suffer great loss which is beyond their capacity.

Figure 1.10

Himalayan glaciers are of different nature than the Arctic glaciers because their highest point is about 3000 meter above the sea level. So these are less influenced by global warming. About 1000 glaciers are present in the Indian region which are melting moreover, the glaciers like Gangotri (Figure 1.11) are decreasing at very high speed. In the coming next 50 years the same may come to an end. If it happens the flow of Ganges river will be stopped. Danger in clearly seen on the important glaciers of Bhagirathi river basin. Contrary to this few glaciers including Siachen glacier (Figure 1.12) are increasing. The world is divided among 16 Geography regions and India represents about 10. So it becomes necessary that India must develop its own system to study the climate changes taking place time to time.

Factors Responsible for Global Warming

1. Increasing Number of Automobiles

In a study report released by Yel university, U.S.A. it was claimed that air pollution in India is highest among the countries of world. According to new report released by EPI, India is the placed 155th among 178 countries of the world the situation in our country is so worse that it stands for behind the countries like Brazil, Russia, China and South Africa. Moreover condition in our neighbouring countries, Pakistan, Sri Lanka and Nepal are much better. Few years back when the air pollution is took a serious stage in China it made many fruitful arrangement in its public transport system, which reduced the air pollution at present the sell of the new cars is restricted to the limit 1.5 Lakh the Amsterdam, (Figures 1.13 and 14) the Capital of Netherlands is now known as the city of bicycles. Where as in our country, only in Delhi about

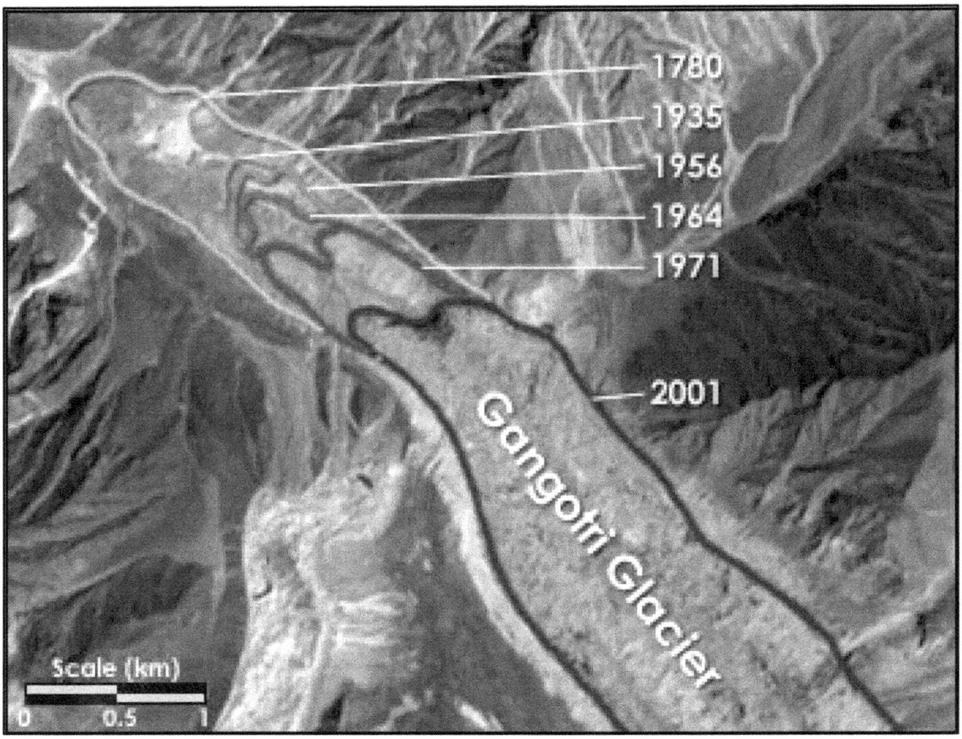

Figure 1.11

Figure 1.12

Figure 1.13

Figure 1.14

1,500 new cars running per day and amazingly there are 8 millions of cars on the road in Delhi (Figures 1.15 and 1.16). We know that automobiles are responsible for release of CO_2, CO, SO_2 gases etc.

Figure 1.15

Figure 1.16

2. Increasing Urbanization

Concentrating our attention on Urbanization, we must admit that the cities of the world as whole utilizes 75 per cent of the total processing and discharge about 50 per cent of the wastages and due to this there is release of 75 per cent GHG.

3. Increasing Industrialization

Industrialization has caused great damage to our environment. Now almost all the water resources like rivers, lakes, ocean, ground water are polluted. Industrial units release affluents many harmful elements which not only effect our environment but these also harm our whole atmosphere. Paper mills release H_2S gas, oil refineries release hydrocarbons and chemical fertilizer faculties release Ammonia. Thermal power plants use cool as a fuel and release Hg and poisonous SO_2 gas. There are about 43 highly polluted industrial areas among which thermal power plant Singroli distt. Sonbhadra, UP is most responsible of atmospheric hazard. Here there are series of private industrial units as well as semi government industries. According to a report 17 per cent of Hg is released in this area which has increased the Hg in blood level of local citizens. Hg is used in chemicals, petro chemical industries, house hold items such as CFL, thermometer etc. The Coal based industries release Hg, which is then spread in soil, dust, water, hence, rivers get polluted. Due to this vegetable, Fishes, sea food become poisonous which then reaches in human body.

4. Ozone Layer Depletion

Ozone layer depletion or reduction in the concentration of Ozone layer allows more ultraviolet radiation to pass through it ultraviolet radiations A [32-220nm], B [220-280nm],C [280-320nm] effects the earth but the B and C are more harmful. Ozone is present in significant amounts in the stratosphere region of atmosphere. Maximum concentration occurs at a height of 15-50 km. (35 km.) from earth level Ozone layer is highly important as it filters harmful, high UV radiation coming from the sun. Ozone depletion is caused by certain chemical called ODS these allows more UV rays to pass through this lays which causes harmful effects on living beings and environment such as–skin cancer, cataract in eyes, damage in immune system, increased embryonic mortality in animals and humans, decreased crop yields and smog formation. In 1985 we found a Ozone hole over Antarctica. Over the years this hole was widened. It has grown in size. Plans are in pipe line to phase out the use of ODS. The use of methyl bromide as a pesticide has already been stopped.

Precaution

Role of UNO in Environment Protection

1. First of all in 3rd December 1968, a resolution was passed in U.N. General assembly and it has proposed to organize a international Conference on Human environment.
2. **Stockholm Summit 1972. (5th June to 16th June)**: A conference was organized by UNO in the important city, Sweden of Stockholm on human environment, in which the topic like air pollution, chemical Toxicity were

discussed globally. In this Conference about 113 countries participated and 5th June was declared as world's Environment day.

3. **Nairobi Summit 1982 :** 105 nation were gathered from 10th to 18 May 1982 in Nairobi and made Nairobi declaration in which these nation announced 10 proclamations (declarations). In the main declaration all government and people of world were urged to perform their duties individually and jointly for protecting the world's environment.
4. **Montreal – Ozone layer protection pact-1987 :** In 1985, a hole was confirmed in the Ozone layer over Antarctica, therefore on 16 Sept, 1987 a Montreal agreement was signed on Ozone protection pact which was declared to come in force from 01.01.1989.
5. **Basel agreement on hazardous waste 1989 :** An agreement was finalized on 22-03-1989 in Basel which was to be applicable from 05.05.1992. According to this agreement no country would throw the hazardous waste in the sea.
6. **Reo earth summit 1992:** In capital of Brazil, Rio de Janeiro, more than 20 thousand representative of 82 Countries participated in this summit. It is known as earth summit; in which all countries have decided to lower down their CO_2 release at 1990 year level similarly.
7. **Kyoto protocol 1997**
8. **Johannesburg earth peak summit, 2002**
9. **Kankun, 2010**
10. **Durban 2011**
11. **Reo+20,2012**
12. **Doha peak summit**
13. Now (Dec. 2014) about 190 nations here participated in 12 days summit held in Leema, a capital of south American city Peru and all nation have decided to reduce their GHG release at the level of 1990.

Note

But we see that till now, nothing is seriously achieved in the matter. Here one important fact is necessary to mention that a polluted environment promotes the criminal activities. Professor E. Bush biken a psychology analyst has described about here change in human behaviour due to environmental circumstances. Pollution works like a trigger in anger. Although devaluation in social, morale and spiritual values are mainly important for crime yet increasing pollution, use of insecticides in edible materials, materialistic way of living, glamorous life style certainly become the reasons to notice. Professor Roger Master of Dortmount college, Hanover in New Hampsire has found a direct connection between environment pollution and crime in America. The cities in which there is excess of Pb and Mn in environment there is the average crime is also double of national level. Due to increase in air pollution there increase the amount of testosterone Hormone, which enhances the cruelty and sex in human being. According to the Gairy Medical College of North Carolina

Hormone serotonin has connection to voilence. Lack of serotonin increases the violence nature of humans and intoxication reduces the amount of serotonin. According to a scientist from Belgium, Prof. Lin Priyara due to the remaining of insecticide DDT, the children's of India and Columbia reach puberty faster. The Remaining of insecticide of Chemical works similar as Harmon Estrogen. Due to the spraying of DDT Mahu and Caterpillar dies which are then eaten by vultures. This effect the Estrogen Harmon of vultures due to which the shells of their eggs become weak and they breaks before time and this leads to foeticide. DDT is the main reason for the decreasing population of vultures and other birds. Dr. William Ambresh a agricultural scientist told that the amount of protein in Wheat and pulses has been decreased by 10 per cent and 2 per cent respectively. Due to the presence of K in chemical fertilizer carbohydrate increases and protein decreases. Famous Doctor Prof. Solomen has told that the food which lack protein increases excitation. In our country pulses are main source of protein for vegetarian.

Results

To overcome the dangers if environmental changes the following steps acts as a remedy:

1. The changes in the weather is effecting the whole world and especially our country as farming is very important in its economy. So we should focus on

Figure 1.17

water harvesting and should use microbiol fertilizers in forms instead of chemical fertilizer because it is cheap, production method is very simple and it produce no pollution problem, they also convert immobilized

Figure 1.18

Figure 1.19

chemical fertilizer into soluble forms and made them available to the plants. We should change the pattern of crops in the case of less rainfall and we should also take care that till 2050 about more than 50 per cent food products would be need due to increase in population so, we have to use genetically modified crops.

2. In the thermal power station the coal which is 50 per cent pollution free should be used as per law of energy conservation. The use of solar and wind energy should be forced because it will reduce the dependence of people on fossil fuel and it will also gradually clean the environment.
3. There is no doubt that the protection of earth environment is very important but its depleting day by due to its over use by humans. There is no such ground in the country which comes 33 per cent upto the mark as per forest area. Law should be strictly imposed for cutting of forest (Figures 1.17–1.19).
4. Funeral is a matter of devotion. But we should take care that for a funeral of a dead body 5 to 6 Quintal of wood for which forest are cutted down. Burning of wood leads to release of GHG now the time has come when we should awaken people to use Electric crematorium.
5. We should also take care that for environmental disbalance consumerist culture is more responsible than increasing population. So we have to change our life style and reduce the use of GHG from industrialisation to transport policies. We should in force the use of public transport instead of private one.

Figure 1.20

Acknowledgement

Authors are thankful to Dr. R.S. Thakur Ex-Director Prabhat Engineering College Kanpur and Dr. N.L. Shukla Principal J.N.P.G. College, Banda (U.P.) and Dr. K.C. Sharma, Principal, Govt. P.G. College, Noida (U.P.) for the suggestions given during the preparation of the manuscript.

References

1. J.P. Mishra, Environmental Chemistry Anusandhan Prakashan, 2012.
2. V.K. Tiwari, Environmental Science, S. Chand, Publication, 2010.
3. Paryavaran Chetna, Editor Meeta Singh. Vol. Dec. 2014.
4. G.C. Pandey, Pratiyogita Darpan, Sep., 2014.
5. Jai Ram Naresh, Danik Jagarn, Kanpur, 7th Oct. 2014; 8th Oct. 2014.
6. O.P. Joshi, Danik Aaj, Kanpur 2nd Feb, 2014.
7. Nagraj Adwe; out look, (U.P.) Dec. 2014.
8. United Nations IPCC. 1st Assessment report Vol. 1 August 1990.
9. IPCC Latest reports.

Chapter 2

Role of Cooperatives in Environmental Protection

Bibhu Santosh Behera[1], Rudra Ashish Behera[2], Anama Charan Behera[3] and Gurudev Sahu[2]

[1]Department of Extension Education, College of Agriculture, OUAT, Bhubaneswar – 751 003, Odisha
[2]Department of MBA, The Techno School, Bhubaneswar, Odisha
[3]D.B. College, Turumunga, Keonjhar, Odisha

ABSTRACT

Cooperatives play a major role in rural areas, particularly where private businesses hesitate to go and public authorities do not provide basic services. They are instrumental in providing opportunities for productive employment, as well as offering health care, education, potable water, improved sanitation, roads, and market access.

The cited paper entitled "Role of Cooperatives in Environmental Protection" narrates the qualitative effort of authors in the realm of development through various Strategies followed by cooperative societies via theoretical analysis.

The issue of environment protection can be left to the government alone. It is the concern of every citizen of the country, cooperatives and its members. As cooperative institutions are peoples, organization, it can play catalytic role in creating awareness among the people. The role could be

1. *Creating awareness about environment protection among cooperatives and cooperative members through extension programmme.*
2. *Developing educational and training programme for Board members and the staff.*
3. *Creating awareness among the people on consequences of felling down tree indiscriminately and shifting cultivation.*

4. *Encouraging people to develop social forestry programme.*
5. *Encouraging people to participate community development programme like rural sanitation and plantation in waste land.*

Keywords: *Cooperative, Environmental Protection.*

Introduction

Cooperatives play a major role in rural areas, particularly where private businesses hesitate to go and public authorities do not provide basic services. They are instrumental in providing opportunities for productive employment, as well as offering health care, education, potable water, improved sanitation, roads, and market access.

Cooperative having emerged as distinct sector is rated as chain agents for the all round socio-economic development through the democratization of its management with a view to harnessing people's creative power. It is known that cooperatives are one of the oldest socio economic organizations in the world. During its long history the cooperative movement has proved itself to be stable, competitive and popular among the people. In fact cooperatives have existed in India since ancient times though in different forms of cooperative organizations.

According to the views of various committees and commissions such as All India Rural Credit Survey Report 1954, National development council's Resolutions 1958, report of the National Commission of Agriculture and various plan documents about cooperative societies, cooperative societies should act as principal agencies of development in economic transformation of villages in India through removing poverty and unemployment. According to Royal Commission on Agriculture (1927), if cooperative fails, there will be the fall of the last hope of rural India.

A cooperative is an enterprise with broader objectives than other corporate forms. It is, "an autonomous association of persons united voluntarily to meet their common economic, social and cultural needs and aspirations through a jointly owned and democratically controlled enterprise."

Cooperatives are based on the values of self-help, democracy, equality, equity and solidarity; as well as ethical values of honesty, openness, social responsibility and caring for others. Cooperatives are guided by a set of underlying values and ethics and are schools of social dialogue and democracy. Cooperatives are often the only provider of services in rural communities, given that other types of enterprises often find it too costly to invest in this area or anticipate low levels of economic return. This is the case for electricity, water resources, financial services, consumer supplies. Cooperative is a philosophy of life and has a mission of its own. It is moral movement; the cooperative spirit helps to make a better persons and a better society. Although cooperative is an organization for the promotion of economic interest of its members, it does not confine itself to the economic aspect of life. It aims at establishing a new democratic social order based on freedom, fraternity, equality and where people live in harmony, caring and sharing like a family. Cooperatives contribute for social development as well as sustainable human development through social integration, education and training, community development, gender equality and protection of against twin evils of rugged individualism and blatant totalitarianism.

Environmental problem and its protection is one of the major concern for survival and welfare of mankind. Due to rapid population growth as well as domestic, commercial and industrial activities this problem becomes more serious particularly in developing countries, like India. Our physical and mental health, custom and culture and very survival are directly and indirectly affected by the environment where we live in. Destruction of environment produces pollution and which results in high environmental hazards to modern civilization.

High population growth contributes a major role in environment pollution. The resources to feed ever-increasing population are very limited. In order to increase food production, chemical fertilizer is used extensively which ultimately degrades the land and leads to various serious ailments of human being. Water resources are going down due to deforestation. Deforestation of forest not only destroys natural habitat, but also increasing run off and increases soil erosion resulting in siltation of river, rivulets and lakes. These create unprecedented flood in the plain. The forest covers are being destroyed to construct buildings and apartment, to get timber and to take out minerals.

Global warming, ozone layer depletion, acidic rain, etc. are the topic of discussion in recent time. Polluted air can cause various diseases and disorder of human eyes, throats and nose. In industrial units release polluted water to nearby rivers and streams without any proper treatment. The consequences environmental pollution has been often ignored and certainly one day we will have to pay a high price for it. Ecologist and environmentalists have been giving warning to the mankind for keeping our environment clean fresh and healthy. A large number of programmes have been initiated by the government and non-governmental organizations (NGOs) for environment protection. At this juncture the role of cooperative cannot be ignored.

Cooperative sector is playing crucial role in sustainable economic development. A cooperative is an enterprise with broader objectives to meet peoples' mutual needs. They are based on powerful idea that a group of people can achieve the economic and social goals of them than individually.

Cooperatives are the organization of the people, farmers, workers and artisans and they have every reason to participate in safeguarding their interests. Because of strong association with member often in specific geographic area, cooperative are also linked to their communities. They have special responsibility to ensure that the development of their community is sustained economically, socially and culturally. Members of cooperatives have responsibility to work steadily for protection of environment where they live in. For this they must evolve the mechanism to integrate environmental concerns into their economic growth and they should work for sustainable development through policies approved by their members. Members should decide in what way the cooperative can make its contribution to the community and preserve the environment. Cooperative institution which operates at the grass root level is the best institution to create awareness among their members and general public as a whole about environment protection. Thus, environmental protection is not confined to a subject alone rather it is movement to save the mankind.

The involvement of cooperatives in protecting environment has been recognized in recent times. The involvement is voluntary and positive, it is noted that consumer cooperative societies in developed countries have taken a leading role in controlling the environment pollution. For instance, use of plastic shopping bags which are non-biodegradable are being gradually reduced and replaced with disposable paper bags.

Agriculture cooperative societies have also made significance contribution for protecting the environment by application of balanced chemical fertilizer, scientific use of insecticides and pesticides, proper deposal agricultural waste.

Cooperatives of different countries are working for environment protection. In Japan, the Japanese Consumer Cooperative Union (JCCU) produces and markets tissue papers from 100 per cent recycled papers. They have committed themselves to conserve environment and sustainable development. Special symbols have been introduced to indicate eco-friendly products put on sale outlet.

In U.K., Cooperative societies have done very good jobs in favour of environment protection. Cooperative retail service limited has cut down on sale of products which are harmful to the environment.

In India, several cooperative organizations mainly fertilizer diary and sugar have responded positively. The national agricultural cooperative marketing federation produces bio-fertilizer and advises its member societies to avoid chemical fertilizer. Indian Farmers Fertiliser Cooperative (IFFCO) has prompted the Indian farm forestry development cooperative Ltd (IFFDC) to regenerate plots of waste land promotes forestation of waste land. The diary and sugar cooperatives have been using solar power for boilers. A green belt has been developed around the existing plant and operates full scale effluent water treatment plant.

Sericulture is an agro based labour intensive industry providing gainful employment opportunities in tribal area through on farm and off farm activities. Sericulture is a livelihood activity that goes around the year and provides remunerative income to farmers. Sericulture development can be achieved through cooperative societies in Odisha. There are three types of silk cultivated in Odisha by three types of cooperative societies such as Primary Tasar Rearers Cooperative Society (TRCS), Mulberry Rearers Cooperative Society (MRCS) and Eri Rearers Cooperative Society (ERCS). Development of sericulture requires heavy plantation of food plant of silk worms. So a large patch of land has been converted to forest for sericulture by rearer cooperative societies. These cooperative societies have positive impact on environment. The cooperative societies working for sericulture development in Odisha protect environment through afforestation for silk worm rearing. Sericulture helps in promoting sustainable development of farmers of rural poor.

Shifting cultivation is otherwise known as slace and burnt cultivation and it is practiced in hilly and forest land. It has serious adverse effect on environment. It destroys valuable biodiversity of the earth. Cooperatives can check the shifting cultivation which affects seriously ecosystem and biodiversity in tribal rural areas by promoting perennial horticultural crops on hilly slopes. The returns of horticultural crops will be much more than traditional crops produced by shifting cultivators.

The issue of environment protection can be left to the government alone. It is the concern of every citizen of the country, cooperatives and its members. As cooperative institutions are peoples, organization, it can play catalytic role in creating awareness among the people. The role could be:

1. Creating awareness about environment protection among cooperatives and cooperative members through extension programmme.
2. Developing educational and training programme for Board members and the staff.
3. Crating awareness among the people on consequences of felling down tree indiscriminately and shifting cultivation.
4. Encouraging people to develop social forestry programme.
5. Encouraging people to participate community development programme like rural sanitation and plantation in waste land.

Though national awareness and planning action is needed, the real work could be done at the grass root level. Cooperative institutions can interact strongly with rural local self government agencies in order to develop an integrated environment development strategy. A well-informed and trained cadre of workers have to develop who could provide education and information on various aspects *e.g.*, aforestation, disposal of garbage, recycling of waste material, rural sanitation, handing chemical fertilizers etc. on a regular basis. The cooperative organization should formulate strategies for environment protection.

The national cooperative and state cooperative training institution should launch intensive programmes of workshop and seminar for education and training of their staff in environmental related activities.

A certain per cent of net profit could be earmarked by each cooperative for environment concern. If considered necessary, statutory provision could be made in this regard in state cooperative laws. This will ensure statutory responsibility on cooperatives and assist cooperatives to lend practical shape to the cooperative principle 'concern for others'.

These are some of the issues which are of prime importance for cooperatives today. Everyone should participate in this programme and cooperatives are best and most appropriate institutions to generate a favourable climate in favour of this movement–a movement to save our mother land and to make it a place to live a healthy life.

References

1. Dutta, S.K. (1991). "Cooperative Societies and Rural Development", Mittal Publications, New Delhi.
2. Nayak, Shipra and Nayak, B. D. (2001). "The role of cooperative societies in socio-economic development of Arunachal Pradesh", *Arunachal University Research Journal* Vol. 4 No. 1.

3. Dharia, M. (2001). "21ST Century and future of Cooperative Movement", Souvenir, State Level Cooerative Seminar on Prospects of cooperative sector in new millennium.

4. Saikia, B. and Jena, S. (2001). "Changing role of marketing cooperatives in Arunachal Pradesh in new millennium", Souvenir, State Level Cooperative Seminar on Prospects of cooperative sector in new millennium.

Chapter 3

Temperature Rise in Jabalpur in Last Thirty Years

Aruna Pande and Rajesh Wahane

Department of Zoology,
Government S.S.A. College, Sihora, M.P.

ABSTRACT

Temperature is an important factor to regulate life processes. Temperature affects plants and animals both. Global warming is a present days problem because its direct and indirect effect on biosphere is harmful. Good health obtained in balanced environmental conditions. The rate of urbanization has shot up in recent years. Recently many Grampanchayat near the Jabalpur city erases their existence. They become the part of Jabalpur Nagar Palika. When city expands it swallows villages. With the villages agricultural land is also become smaller and smaller. Most of the agricultural land is diverted for building making purpose. Builders use diverted agriculture land to build residential complexes. The crop production has ceased and in place of green fluttering harvest stable concrete structures are coming up. This become the cause of environment warming and creates health problems. Agriculture cools the environment by irrigation, transpiration, absorption and use of carbon dioxide with sunlight for food production. In Jabalpur temperature differences from 1978 to 2008 is observed. The numbers of building projects are also increasing. Agricultural land of peripheral villages had converted for these purposes. The rise in temperature and diversion and shortening of agricultural land may correlate. Increasing temperature harms the human as well as other animal's health directly or indirectly. Crop balances the carbon dioxide concentration in atmosphere. The carbon dioxide concentration affects the nutritive value of crops. With the carbon dioxide concentration, the quantity of certain elements in the crops varies. This affects the health of human beings. So the reduction of agricultural land may be harmful to human health by changing the amount of carbon dioxide and increasing environmental temperature.

Keywords: *Agricultural land, Diversion, Carbon dioxide concentration and Temperature.*

Introduction

Diversion of agricultural land for building purpose is continuously increasing. Agriculture not only fulfils food problems but also gives an environment suitable for biosphere. Control on soil erosion and maintenance of underground water level are side products of agriculture. With this agriculture also lower the atmosphere temperature by many ways. In this study an observation is made to know the temperature change in Jabalpur city, where many small villages are going to disappeared and expansion of city is continue.

Materials and Method

Many building projects are running in villages around Jabalpur city. The minimum and maximum temperature of Jabalpur from 1978 to 1982 and after thirty years that is from 2008 to 2012 is taken from India Meteorological Department Jabalpur. We analyze the temperatures of both periods to know about the temperature change after thirty years.

Observation and Results

There is many villages near Jabalpur city where building projects on diverted agricultural land are going on. Some are Tilhari, Bilhari, Barela, Amkhera, Kajarwara, Rimjha, Neemkheda, Paderiya, Lamti etc. Many Grampanchayats merged in Jabalpur Nagar Nigam. Due to this the area of city expanded and Seventy seated Nagar Nigam become Seventy Nine seated.

We took the minimum and maximum temperature in two group from India Meteorological Department Jabalpur. One group is of five years from 1978 to 1982 and another is also of five years from 2008 to 2012. Actually in last ten years most of the building projects are started and achieved heights (Tables 3.1 and 3.2).

Table 3.1: Average Minimum Temperature (°C) for Year 1978–1982 and 2008–2012

Month	*Year*									
	1978	*1979*	*1980*	*1981*	*1982*	*2008*	*2009*	*2010*	*2011*	*2012*
JANUARY	8.89	11.86	11.2	10.15	13.21	10.8	13.2	8.8	8.6	10.8
FEBRUARY	13.06	11.72	12.64	12.7	13.12	10.7	14.3	14.1	13.9	16.7
MARCH	15.93	15.29	16.99	16.64	15.05	19.3	18.5	20.3	19.3	26.3
APRIL	21.23	21.94	23.55	21.07	21.7	22.7	22.5	25.5	23.7	24
MAY	27.41	24.29	28.97	26.53	24.04	27.3	27.9	29.2	28.3	26.8
JUNE	25.56	26.49	25.97	28.14	26	25	27.9	27.8	25.7	27.7
JULY	24	24.64	24.13	24.18	24.72	24.7	24.9	25.3	24.6	24.6
AUGUST	23.4	23.49	22.52	24.9	23.3	24.4	24.4	24.6	24.3	24
SEPTEMBER	22.97	23.31	23.22	22.31	21.07	24	24.2	23.5	23.7	24.1
OCTOBER	17.95	19.97	19.72	19.32	20.3	19.7	25.5	21.3	19.9	19.6
NOVEMBER	15.73	17.87	14.4	12.98	15.37	15	15.8	18.6	15.3	14.3
DECEMBER	9.58	11.41	10.6	9.28	12	12	17.1	11.1	10.1	14.3
AVERAGE	18.8	19.3	19.4	19.1	19.1	19.6	21.3	20.8	19.7	21.1

Table 3.2: Average Maximum Temperature (°C) for Year 1978–1982 and 2008–2012

Month	*Year*									
	1978	*1979*	*1980*	*1981*	*1982*	*2008*	*2009*	*2010*	*2011*	*2012*
JANUARY	23.55	24.89	26.57	23.76	25.71	23.8	23.8	23.2	24.2	22.1
FEBRUARY	25.01	25.55	29.93	28.41	25.96	26	29.6	28.1	28.3	27.3
MARCH	30.64	32.25	34.16	32.57	31.29	38.9	34.4	35.9	34	33.4
APRIL	37.93	38.89	40.65	39.16	37.15	38.6	38.6	40.9	37.3	38.7
MAY	42.6	40.08	43.11	40.68	38.89	40.8	41.8	42.7	40.1	41.4
JUNE	35.49	39.57	36.5	39.54	38.11	34.1	41.2	39.5	34.7	38.3
JULY	30.06	32.15	30.23	32.47	33.51	31.2	31.6	32.4	30.9	30.9
AUGUST	26.75	30.78	31.5	31.11	28.28	30.6	31.5	31.1	30.1	28.6
SEPTEMBER	31.25	33.7	31.36	32.97	31.85	31.1	31.9	30.7	30.4	31.3
OCTOBER	31.34	34.76	33.59	32.12	31.89	32.6	31.7	31.5	31.4	32
NOVEMBER	29.6	31.72	29.67	28.11	27.8	28.3	26.8	29.7	29.9	28
DECEMBER	23.4	23.03	25.74	24.35	27.05	27.3	26.7	25.1	25.6	26.6
AVERAGE	30.6	32.2	32.7	32.1	31.4	31.9	32.4	32.5	31.4	31.5

Increase in Minimum Temperature

We analyzed the average temperature in different months of the different year. The average minimum temperature is ranged between 8°C to 30°C. The average minimum temperature of cool months like January and December is up to 10°C. The temperature of May and June is above 25°C. The average minimum temperature of April months from 1978 to 1982 was 21.2°C to 21.94°C except in April 1980 it was 23.55°C. whereas the average minimum temperature of April month from 2008 to 2012 was 22.5°C to 25.5°C. Like this the average minimum temperature of September month from 1978 to 1982 was 21.7°C to 23.3°C and from 2008 to 2012 it was from 23.5°C to 24.2°C. It was observed that in sixty months from 1978 to 1982 the average minimum temperature was above 25°C in eight months and from 2008 to 2010 it was in thirteen months (Table 3.1 and 3.3).

When we sorted the months according to 1°C temperature range from 24°C to 29.9°C, we found that within 1978 to 1982 averages minimum temperature was above 24°C in sixteen months where as it was in twenty seven months from 2008 to 2012. Only in three months from 1978 to 1982 the average minimum temperature was more than 27°C but in seven months from 2008 to 2012. This shows slight increase in minimum temperature from 1978 to 2012 (Table 3.4).

Increase in Maximum Temperature

(a) Average maximum temperature of different month is ranges between 23°C to 43°C. In the month of February in 1978 to 1982 the average maximum temperature was 25°C to 29°C, whereas it was between 26°C to 29°C in the February months of 2008 to 2012. Like this the average maximum temperature of March months was 30°C to 34°C from 1978 to 1982 and it was between 33°C to 38°C in 2008 to 2012 (Table 3.2).

Table 3.3: Average Minimum Temperature (5°C) in Number of Months from Year 1978–1982 and 2008–2012

Temperature	*Year*											
	1978	*1979*	*1980*	*1981*	*1982*	*Total*	*2008*	*2009*	*2010*	*2011*	*2012*	*Total*
Up To 9.9°C	2	0	0	1	0	3	0	0	1	1	0	2
10.0°C to 14.9°C	1	3	4	3	3	14	3	2	2	2	3	12
15.0°C to 19.9°C	3	3	2	2	2	12	3	3	1	3	2	12
20.0°C to 24.9°C	4	5	4	4	6	23	4	4	4	4	4	20
25.0°C to 29.9°C	2	1	2	2	1	8	2	3	4	2	3	14

Table 3.4: Comparison of Average Minimum Temperature (1°C) in Number of Months from year 1978–1982 and 2008–2012

Year	*Temperature*						
	24.0°C to 24.9°C	*25.0°C to 25.9°C*	*26.0°C to 26.9°C*	*27.0°C to 27.9°C*	*28.0°C to 28.9°C*	*29.0°C to 29.9°C*	*Total*
1978 to 1982	8	2	3	1	2	–	16
2008 to 2012	13	5	2	5	1	1	27

Table 3.5: Average Maximum Temperature (5°C) in Number of Months from year 1978–1982 and 2008–2012

Temperature	*Year*											
	1978	*1979*	*1980*	*1981*	*1982*	*Total*	*2008*	*2009*	*2010*	*2011*	*2012*	*Total*
Up to 24.0°C	2	2	0	2	0	6	1	1	1	1	1	5
25.0°C to 29.9°C	3	1	4	2	5	15	3	3	3	3	4	16
30.0°C to 34.9°C	4	6	5	5	4	24	5	5	4	6	4	24
35.0°C to 39.9°C	2	2	1	2	3	10	2	1	2	1	2	8
MORE THAN 40.0°C	1	1	2	1	0	5	1	1	2	2	1	7

More than 40 average maximum temperatures was observed in five months out of sixty months from 1978 to 1982, but it was in seven months between 2008 to 2012 (Table 3.5).

We observed 2 to 5°C more temperature on the main roads of Sihora then the farms and gardens.

Discussion

Increasing global warming is a current issue, which influencing biosphere very badly. Global warming is the result of changed climatic and ecological conditions. Human health terrestrial and aquatic ecological systems, agriculture, forestry, fisheries and water resources are sensitive to change in climate. In nature each biological and physical unit is capable to affect each other. To days the thought that urbanization is main cause of global warming is strengthened by many scientists.

There have been suggestions that a proportion of the 0.5°C warming seen over the last century may be related to urbanization influences

The impact of urbanization in eastern china on surface warming is analyzed. Result show that the urbanization can induce a remarkable summer warming in Yangtze River Delta (YRD) city cluster region and a winter warming in Beijing-Tianjn Hebel (BTH) city cluster region. The YRD warming in summer primarily results from the significant increasing of maximum temperature. The BTH warming in winter is primarily due to remarkable increasing of minimum temperature.

In our study we observe increase in minimum temperature in eleven month in Jabalpur city. The process of urbanization is a change of rural area in to urban area. In this change the agricultural lands are diverted for the building purpose. The results of agricultural land diversion is like the effects of deforestation.

A micro-scale example of climate change is already occurring in metropolitan areas. In advertent weather modification in urban areas from important 'lesions' for global climate change, as noted by Changron.

Urban development has influenced the temperature records of even small rural cities in Turkey, in a way that is compatible with result of other studies conducted in North America.

Agriculture cools the environment many ways. The temperature near the crops is decreases by irrigation. Transpiration spreads water vapors in the environment and helping in down the temperature. Plants used sun light for photosynthesis and in this way they lowers the environmental temperature. Plants manufacture food from Caron dioxide and control the green house gas Carbon dioxide. By this they helps in lowering down the temperature.

Amongst the green house gases Carbon dioxide is the most important accounting for 60 per cent of global warming. Carbon emission is above the earth's carbon fixing capacity. Plants play an important role in carbon balancing.

The world's soils hold about twice (1400–1500 Gt C) as much carbon as the atmosphere. Carbon stored in agricultural soil is 170 Gt, while the entire vegetation contains 550 Gt C (Paustian), Emission of carbon dioxide due to deforestation is considered to be a major source of atmospheric carbon dioxide. In India the amount of carbon stored in the soil is 23.4- 27.1 Gt, which is 1.6 to 1.8 per cent of the carbon stored in the world's soils.

Agriculture may be used to maintain suitable environmental conditions. The earliest archeological evidences found throughout the tropical and subtropical areas of southeastern and southern Asia, northern and Central Africa and Central America, suggest rapid and large scale domestication of plants and animals ca. 10000-7000 cal years BP. This interval corresponds to an intense humid phase and equable climates, as observed in numerous paleo records across the regions. Gupta suggested that domestication of plants and animals and subsequent beginning of agriculture were linked to climate amelioration in early Holocene.

On May 22 Chicago Council on Global Affairs released a report urging the US government to take action to curb the risks the climatic change poses to global food security. It explains how higher temperature changes in rainfall and natural disasters caused by climate change could undermine food production and put food supplies at risk. In total climate change could reduce food production growth by 2 per cent each decade for the rest of this century

The report was released at the Chicago Council's on Global Food Security Symposium 2014. Reduced food production will create the health problems. According to Advancing Global Food Security in the face of changing climate 2014 today more than 840 million people are chronically hungry. It is predicted that by 2050 population growth rising increase and changing diets will increase food demand by 60 per cent.

Yet rising temperature changes in precipitation and increasing weather extremes will impact the food system. Changes are already evident in the United States and other countries and will only accelerate in the future.

When grassland and forests are converted to agriculture the loss of vegetation above ground results in large carbon dioxide emissions. This land use change accounts for about 6 per cent of the warming potential from total green house gas emission annually mainly through deforestation in tropical regions

In general as average temperature increases to between 68° F (20°C) (for crops that evolved in temperate regions such as Wheat) and 86° F (30°C) (for crops that evolved in tropical regions such as Cassava and rice) growth accelerates for plants. Above 88° F (31°C) the productivity of most crops drops rapidly. Especially when combined with moisture deficits. Increases in maximum temperatures can lead to serve yield reductions and reproductive failure in many crops.

Increased night time temperatures reduce rice yields by up to 10 per cent for each 1.8° F (1°C) increase in current minimum temperature in the dry season in the humid tropics.

Vegetables are generally sensitive to environmental extremes especially high temperature.

Studies published between 1993 and 2003 have attributed 0.7–3.6 per cent increase in all causes of mortality for each °C increase in temperature above a (local) threshold in several European cities.

Conclusion

The agriculture not gives only food it maintains the climatic condition also. The increase in temperature is not good for health. To control the rise in temperature, the diversion of agricultural land should be minimized.

References

1. Kukla, G., Gavin, J. and Karl, T.R., 1986 ."Urban Warming". *J. Clim. Appl. Meteorol,* 25, 1265-1270.
2. Wood, F.B.; 1988. Comment on the need for validation of the Jones *et al.,* "Temperature trends with respect to urban warming". *Clim. Change* 12, 292-312.
3. Wukai and Yang Xiuqun: 2013. "Urbanization and heterogenous surface warming in eastern china". *Chinese Science Bulletin.* 58/12, 1363-1673.
4. Changnon, S.A.; 1992. "Inadvertent weather modification in urban areas, Lessons for global climate change", *Bull. Am. Meteorol.* Soc. 73, 619-627.
5. Gupta, R.K. and Rao, D.L.N. *Current Science,* 1994, 66-73.
6. Robertson, G.P. and Tiedje, J.M., *Soil Boil. Biochem.* 1987, 19, 187-193.
7. Luizao, F., Matson, P., Livingston, G., Luizao, R. and Vitousek, P., *Global Biogeochem cycle,* 1989, 3, 281-285.
8. Anil, K. Gupta: 2004. Origin of agriculture and domestication of plants and animals linked to early Holocene climate amelioration. *Current Science,* vol. 87, No. 1, 10, pp. 54-59.
9. Christopher B. Field *et al..* "Summary for policy makers", in climate change 2014, Geneva: IPCC Secretariat 2014.
10. Jerry M., Melillo, Terese T.C., Richmond, Gary W., Yohe, eds, climate change impacts in the United States: The Third National Assessment 2014.
11. Sonja J., Vereulen, Bruce M., Campbell and John S.I., Ingram, 2012. "Climate change and food systems", *Annual Review of Environmental Resources* 37: 195–222.
12. Philip K., Thornton and Laura Cramer eds., 2012. Impacts of climate change on the agriculture and aquatic systems and natural resources within the CGIAR Working Paper No. 23.
13. Suzanna K., Redfern, Nadina Azzu and Jesie S. and Binamira, 2012. " *Rice in South East Asia.*
14. Thornton and Cramer, eds., Impacts of climate change on the agricultural and aquatic systems and natural resources with in the CGIAR Mandate, 2012.

15. Kunset A., Looman, Mackenbach J. Outdoor air temperature and mortality in the Netherlands; a time series analyses, *Am. J. Epidemiol.* 1993, 137, 331-341.

16. Pattenden S., Nikiforov B., Armstrog B.G., Mortality and temperature in Sofia and London *J. epidemiol. Community health* 2003; 57, 628-633.

Chapter 4
Role of NGOs in Nature Conservation

Dilip N. Pandit
Alma Times, Indore, Madhya Pradesh

Introduction

Globalization has considerably weakened traditional governance processes. Increasing global economic integration has reduced the power of national governments while granting other economic and political actors access to the world stage. The 1990s witnessed a dramatic increase in the involvement of non-governmental organizations (NGOs) in global governance (Charnovitz, 1997). NGOs and other civil society groups are not only stakeholders in governance, but also a driving force behind greater international cooperation through the active mobilization of public support for international agreements. Enabling the constructive participation of civil society in global environmental governance is thus one of the most important tasks for policymakers concerned with the effectiveness of global governance (Gemmill, Ivanova, and Chee, 2002).

Dozens of Governments, businesses, civil society and indigenous peoples participated in the United Nations Climate Summit in New York pledged to halve deforestation by 2020 and to end within the following decade. "Forests are not only a critical part of the climate solution – they hold multiple benefits for all members of society," Secretary-General Ban Ki-moon said in an announcement about the initiative. The New York Declaration on Forests, a non-legally binding political agreement, calls for the restoration of more than 350 million hectares of forests and croplands, an area greater than the size of India. Deforestation is a significant contributor to climate change. Trees, which store carbon, release it when they are burned during slash-and-burn land clearing of forests, for example. If it works as expected, the initiative would avoid between 4.5 and 8.8 billion tons of carbon dioxide each year by 2030. That is equivalent to removing the carbon emissions produced by the one billion cars that are currently on the world's roads.

"The New York Declaration aims to reduce more climate pollution each year than the United States emits annually," In support of the Declaration, 20 global food companies, most recently Dunkin Donuts and Krispy Kreme, announced their pledges to deforestation-free sourcing policies of palm oil. The world's largest palm oil companies – Wilmar, Golden Agri-Resources and Cargill – also committed to work together to implement and join the Indonesian Business Council in asking incoming Indonesian President Joko Widodo to support their efforts through legislation and policies.

Taken together, the share of palm oil under zero deforestation commitments has grown from 0 to about 60 per cent in the last year, with the potential to reduce up to 450 million tons of carbon dioxide emissions annually by 2020. That is the equivalent of 2 billion tonnes in the period through 2020. Among other announcements, 26 governors from Peru and Liberia presented new forest policies and pledged to cut deforestation by 80 per cent.

Meanwhile, the Democratic Republic of the Congo (DRC), Ethiopia, Guatemala, Uganda and several other countries are set to make national pledges to restore more than 30 million hectares of degraded lands. Also, the Consumer Goods Forum, a coalition of 400 companies, called on Governments to pass a legally binding climate deal in Paris in 2015 that includes REDD+, including large-scale payments to countries that reduce deforestation. More than 120 Heads of States and Government, business, finance and civil society representatives are expected to announce commitments that will reduce emissions, enhance resistance to climate change and mobilize financing for climate action.

India is a party to the UN Convention to Combat Desertification (UNCCD) and MoEF is the National Coordinating Agency for the implementation of the UNCCD in the country. As an affected party, a 20 years comprehensive National Action Programme (NAP) to combat desertification in the country has been prepared. The objectives are:

1. Community based approach to development
2. Activities to improve the quality of life of the local communities
3. Awareness raising
4. Drought management preparedness and mitigation
5. R and D initiatives and interventions which are locally suited
6. Strengthening self governance leading to empowerment of local communities.

Environment and Role of NGOs

There is a wide spectrum of Environmental NGOs categorizable in terms of their specific fields, levels of operation and the extent to which they work to influence government. There are some groups which lobby for more stringent legislation and work to influence and inform public officials. Other groups specialize in research, training, public education, information dissemination, acquisition and management of protected areas, or convening of stakeholders in environmental conflicts. There are groups involved in urban-based educational programmes while others focus on 'hands-on' rural development at the grassroots level. NGOs play vital role in various activities as:

Information collection and dissemination;

- ✰ Policy development consultation;
- ✰ Policy implementation;
- ✰ Assessment and monitoring;
- ✰ Advocacy for environmental justice.

Role of NGOs and Environment and Nature Conservation

According to Govt. of India there are more than 22 lacs registered NGOs. There is many kinds of nongovernmental organizations (NGOs) engaged in working towards the improvement of the social, economic, environmental and political conditions of the country. They include citizens' associations, professional societies, foundations, some universities and research centers, and networks of experts or institutions. There is a wide spectrum of Environmental NGOs categories in terms of their specific fields, levels of operation and the extent to which they work to influence government. There are some groups which lobby for more stringent legislation and work to influence and inform public officials. Other groups specialize in research, training, public education, information dissemination, acquisition and management of protected areas, or convening of stakeholders in environmental conflicts. There are groups involved in urban-based educational programmes while others focus on 'hands-on' rural development at the grassroots level.

WWF India was the first NGO in India to promote nature conservation on a nationwide basis at a time when this was regarded as an eccentric elitist whim. Through its Conservation Action Support Programme (CASP), WWF India supports grass-roots level activities taken up by the small NGOs. Since 1984, WWF India has been a part of the Environmental Information System (ENVIS) of the Ministry of Environment and Forests. The Directory of Environmental NGOs in India, now in its 9th edition, is a periodic publication of the NGO database maintained by the ENVIS Centre. This first of its kind, comprehensive information bank takes into account existing NGOs in all the States and Union Territories working towards environmental protection, conservation and awareness. It provides critical information about the groups engaged in the cultivation and nurturing of interest in environmental protection and the conservation of nature and natural resources.

The spectrum of 'Missions' as provided by the different listed NGOs ranges from:

- ✰ Promoting environmental education
- ✰ Conducting awareness camps

- Safeguarding environment through social mobilization and community participation
- Biodiversity and Wildlife conservation
- Various other research activities and movements

In 1984, ENVIS Centre 07 of the Ministry of Environment and Forests, Government of India, was established in WWF-India to handle environment-related information on the Parliament, NGOs and the media. The Centre has been actively monitoring parliamentary discussions on environmental issues. A substantial amount of environment-related information is exchanged in both the Houses of Parliament in response to queries and call attention notices raised by the Members. Such information relates largely to development activities with environmental implications, other relevant programmes, legislation, policies, etc. This information has been compiled and published by the Centre in separate volumes on a yearly basis. In each volume, questions/answers and statements are presented verbatim from the replies of Ministries on environment matters. The data is procured from the Parliament Secretariat, New Delhi as well as from the website of the Indian Parliament. Beginning from 2005, the Centre has enhanced the analytical section by adding several new features. One addition pertains to highlighting the participation of each political party through the number of 'Green' questions that have been tabled by the Members. Also, the state-wise distribution of questions has been prepared subject-wise to present a broad overview of the environmental problems prevalent in the states and union territories. The section of statistical and graphical representation is an important source of ready assessment of the trend of environment-related information being transacted annually in the Parliament. Each question/answer and statement has been classified under selected subject heads listed under 18 (Eighteen) categories. The questions put in the Parliament are categorized according to these subjects. The list (showing the coverage areas of the subjects) is given below:

A significant growth of NGOs started after India achieved independence. Democracy was established and people had started understanding the meaning of freedom of speech, the charm of equality and the value of humanity and brotherhood. Also, on the other hand, the government started planning for development and in this effort, launched inter alia the schemes of Community Development Programme and later on the Green Revolution.

Over one million NGOs are pursuing their economically gainful activities. The achievements of the schemes were assessed were found to be unsuccessful in providing for the minimum necessary requirements and reducing the gap between the rich and the poor in rural areas, instead rather increasing it.

With independence also accelerated the processes of industrialization, urbanization, expansion of education, politicization and democratization and modernization. These processes awakened people to be sensitive to the existing inequalities like economic inequalities (landed inequality in particular), gender inequality, inhuman kind of social segregations like caste inequalities and untouchability, other social evils like child marriage, child labour, restriction on widow marriage and many other stigmas and taboos.

Industrialization and urbanization led to the emergence of the problems of rural-urban migration causing imbalance in the rural division of labour, over-urbanization of many cities with the expansion of slum and pavement dwellers, urban unemployment, pollution and depletion of natural resources.

NGOs have gained importance now and are increasing in number very fast. Enhancement of their importance is the result of weakening of the role of state in upholding the welfare and well-being of its citizens and consequent development of the assertive role of civil society to ascertain social welfare and integration. More than half a million voluntary organizations would perhaps be working in the country.

Role of NGOs in Development

NGOs have immense role in bringing about social change and development and it is being experienced from different parts of the country. Development, as we have read earlier, is a multi-faceted process, which essentially involves the aggressive participation of the people that would not be possible unless they are educated, awakened and motivated. NGOs are taking up this job sportingly and successfully.

The NGOs are active to promote education, particularly among that section of population, which has remained un-benefited or less benefited by the measures adopted by the government. The education of girls, and other deprived people, particularly the SCs and STs, has been their target objective. Women are the other vulnerable section of society. Gender discrimination is a ubiquitous cultural reality. Girls are discriminated in the upbringing pattern in the family. Larger numbers of the undernourished are from amongst the girls. Retention of girls in schools is much less as compared to boys. Women are forced to work as housewife and denied participation in gainful economic activities outside homes. About three-fourths of the work done by women is un-monetized.

The role of women voluntary organizations towards this cause has been marvelous. Sewa, Sathin, Eklavya, Disha, Environmental Action Group and Agrani Foundation etc. are some of the thousands of NGOs known for their role in development by creating awareness among people and interventions, if required. The approach to development has been almost uniform world over at least in terms of the use of technology, magnitude of production, pattern of consumption and achievement of wealth. Both state and people were unaware or lackadaisical about the backwash of the nature of development pursued. The threat to the human life developed due to environmental pollution and imbalance and the depletion of natural resources as a consequence of the nature of development. Here, the role of NGOs is really noticeable and praiseworthy. Thousands of voluntary organizations are at work to awaken people and governments against environmental degradation and depletion of resources. It is not that the development process has unleashed only environmental threats to the human existence but also many people are displaced due to developmental projects and are quite often not properly compensated and rehabilitated. The NGOs have a major role to play towards the cause of people's resettlement and are also performing commendable job in this direction. The projects like the construction of dams, road highways and railways have often made some sections of people, particularly in rural areas, vulnerable and are displaced without being properly compensated.

NGOs are also rendering great service in restoring dignity to the deprived and discriminated sections of the people in the society like women suffering from gender discrimination, lower caste people suffering from caste segregation and the status of untouchable, racial and religious discriminations. Voluntary organizations, working at both national and international levels, have earned praise for their services in societal development. These organizations are busy in creating awareness and zeal for participation in development projects. Ensuring humanism by fighting against human rights violations, social exclusions, domestic violence and others have been common objectives of the NGOs. Of late, these organizations are also entering the sector of economic well-being and standard of living. In the state of Uttar Pradesh, Agrani Foundation's Jan Suraksha Kranti (JSK) scheme of savings and life insurance is indeed a commendable effort in this direction.

References

1. Abahlali base Mjondolo *Rethinking Public Participation from Below*, 'Critical Dialogue', 2006.
2. Cox, P. N. Shams, G. C. Jahn, P. Erickson and P. Hicks. 2002. Building collaboration between NGOs and agricultural research iNGOs – Die Gewerkschaften in Guinea während der Unruhen 2007] – *EPU Research Papers: Issue 03/07, Stadtschlaining 2007 (German).*
3. Davies, T. 2014. NGOs: A New History of Transnational Civil Society. New York: Oxford University Press. ISBN 9780199387533.
4. http://envfor.nic.in/about-ministry/envis-centres
5. http://www.indiaenvironmentportal.org.in/directory/?page=3
6. http://www.wwfenvis.nic.in/Content/Parliament_3240.aspx
7. Lyal S. Sunga, "Dilemmas facing INGOs in coalition-occupied Iraq", in Ethics in Action: The Ethical Challenges of International Human Rights Nongovernmental Organizations, edited by Daniel A. Bell and Jean-Marc Coicaud, Cambridge Univ. and United Nations Univ. Press, 2007.
8. Lyal S. Sunga, "NGO Involvement in International Human Rights Monitoring, International Human Rights Law and Non-Governmental Organizations" (2005) 41-69.
9. Mark Butler, with Thulani Ndlazi, David Ntseng, Graham Philpott, and Nomusa Sokhela. *NGO Practice and the Possibility of Freedom* Church Land Programme, Pietermaritzburg, South Africa 2007 Churchland.co.za
10. Olivier Berthoud, *NGOs: Somewhere between Compassion, Profitability and Solidarity* Envio.org.ni, PDF Edinter.net Envio, Managua, 2001.
11. Steve Charnovitz, "Two Centuries of Participation: NGOs and International Governance," Michigan Journal of International Law, Vol. 18, Winter 1997, at 183-286.

12. Steve W. Witt, ed. Changing Roles of NGOs in the Creation, Storage, and Dissemination of Information in Developing Countries (Saur, 2006). ISBN 3-598-22030-8.
13. Susan Cotts Watkins, Ann Swidler, and Thomas Hannan. 2012. "Outsourcing Social Transformation: Development NGOs as Organizations." Annual Review of Sociology, Vol. 38, pp. 285–315.
14. Terje Tvedt, 19982/2003: Angels of Mercy or Development Diplomats. NGOs and Foreign Aid, Oxford: James Currey.
15. The Role of NGOs and Civil Society in Global Environmental Governance by Barbara Gemmill and Abimbola Bamidele-Izu.
16. Velusamy M. Non Governmental Organisation, Dominant Publishers and Distribution Ltd, New Delhi.
17. Werker and Ahmed (2008): What do Non-Governmental Organizations do?

Chapter 5

Enhancing Community Based Food and Water Security through Micro-Diversion Based Irrigation Initiative in Poverty Stricken Areas of Orissa

Bibhu Santosh Behera[1], Rudra Ashish Behera[2] and Anama Charan Behera[3]

[1]Department of Extension Education, College of Agriculture, OUAT, Bhubaneswar – 751 003, Odisha
[2]Department of MBA, The Techno School, Bhubaneswar, Odisha
[3]D.B. Junior College, Turumunga, Keonjhar, Odisha

ABSTRACT

The present chapter entitled as" Enhancing Community based Food and Water Security through Micro-Diversion Based Irrigation Initiative in poverty stricken areas of Orissa" was undertaken in Nayagarh District, a tribal district having hilly areas with major irrigation problems. In 7 GPs the DBI structures had been developed via self help mode and assistance from SDTT-Udyama convulgence. The goal of this project seeks to further lasting improvement in quality of life and a tangible reduction in distress migration of poor families and economic excludes through enabling local community empowerment. This project will acknowledge the enhancement of social justice where in house hold livelihood security, food security; entitlements to assured employment along with addressing the emerging issues like human adaptation are ensured and maintained. This project seeks to address the inherent problems associated with human living and gross nature of social intervention applicable for deprived sections of the society. Here life saving wells were built up for water conservation and preservation for future. With capillarity system of pipes and gravitational force the water will flow and supply to farmer for both agriculture and daily use. The major objectives of the study were as follows.

Objective Activities (with a time plan)

- ☆ Construction of community based micro-diversion gravity flow irrigation Infrastructures in a- risk villages.
- ☆ Protection of catchments through intensive conservation practices (mechanical, technical and biological measures with social measures) based on local available resources with minimal external inputs.
- ☆ Promotion of farming groups/user groups for creating opportunity to utilise the harvested water in command area for enhancing production through SRI, and allied agriculture practices.
- ☆ Appropriate and adopted measures on soil and water conservation would be under taken in order to reduce external support dependency.
- ☆ Demystification of technology on soil, water, and animal and forest management as part of farm based livelihood diversification support system.
- ☆ Restoring traditional land and water harvesting/management systems and creating new ones where required.

Expected Output/Impact

- ☆ The project will provide food and water security to approx. 2180 population of 07 villages.
- ☆ Appx. 95 hectares of land in kharif and 65 hectares of land in *rabi* season will be developed and agricultural, SRI horticulture and plantation will be promoted.
- ☆ 70 farmer both male and female will be trained on organic farming and SRI.
- ☆ 70 farmers will be exposed to the other agricultural best practice region.
- ☆ 07 WSHGs and 07 farmers club will be promoted with enterprise development and the skill of the women will be developing.
- ☆ 07 greenery spot will developed.
- ☆ 07 nos. of community safety nets on water, grain banks and energy—bio-mass security will be promoted.
- ☆ Substantial distress migration and drudgery will be reduced.
- ☆ **Location of work**
- ☆ 7 poverty stricken Grampanchayats of Nayagarh Dist. are included in this project.
- ☆ The proposed DBI project villages are from Nuagaon and Dasapalla block of Nayagarh District.
- ☆ Details on districts, DBI structures and cost.

Details of proposed DBI Villages

Sl.No.	*District*	*Block*	*Village*	*Appx. Population Benefited*	*Area to be Irrigated (Acre)*
1.	Nayagarh	Nuagaon	Chakapathar	350	35
2.	Nayagarh	Nuagaon	Chadei mara	260	30
3.	Nayagarh	Nuagaon	Badakathabari	235	31
4.	Nayagarh	Nuagarh	Khalmada	450	60
5.	Nayagarh	Dasapalla	Barapalli	310	32
6.	Nayagarh	Dasapalla	Satajhari	275	25
7.	Nayagarh	Dasapalla	Durgaprasad	300	23

Project Target

The project would target to empower vulnerable section to access potential schemes in the locality. The schemes would very well be government and non-government. The issues of the vulnerable groups would be analyzed and correlated to potential schemes and training would follow for the groups on the select schemes. Essentially the vulnerable groups will be decided in community consultation process and hence it would not suffer from Caste/Gender or Religious biases. The special emphasis would be given on intensive follow up with ST, SC and Women groups to hand hold to access the schemes.

Research Methodology

Here the Methodology is followed as below:

1. Learning by doing method and seeing is believing.
2. Data Collection by Participatory Rural Appraisal Method and its components.
3. Problem Identification: By SWOT, SMART analysis and Problem Tree Analysis method through community based focused meeting.

Figure 5.1

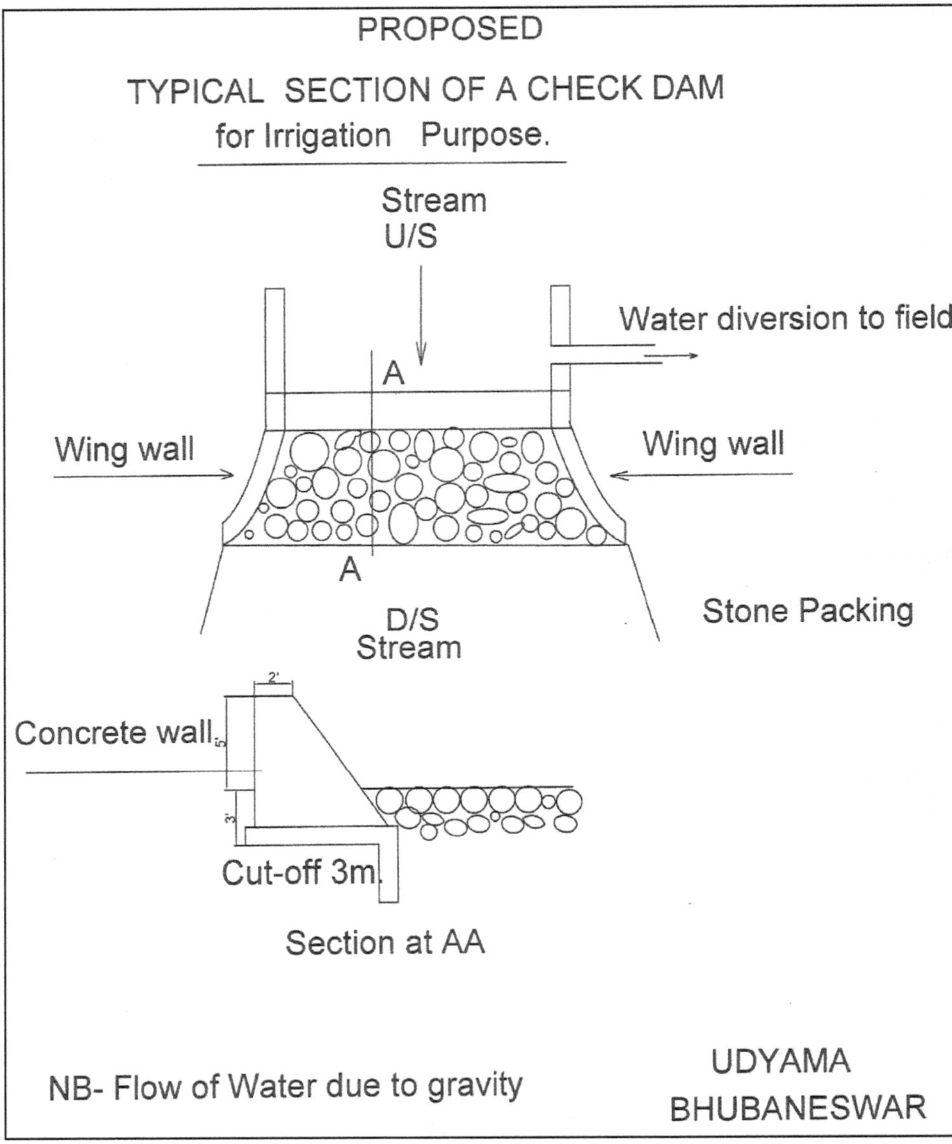

Figure 5.2

4. Models have been used as per locality and developed by using Auto-CAD method.
5. Research design: Post facto design using purposive sampling.
6. Instrument used for data collection: Structured interview schedule with open ended format.
7. Research statistics used: Rank order, mean, and percentage.

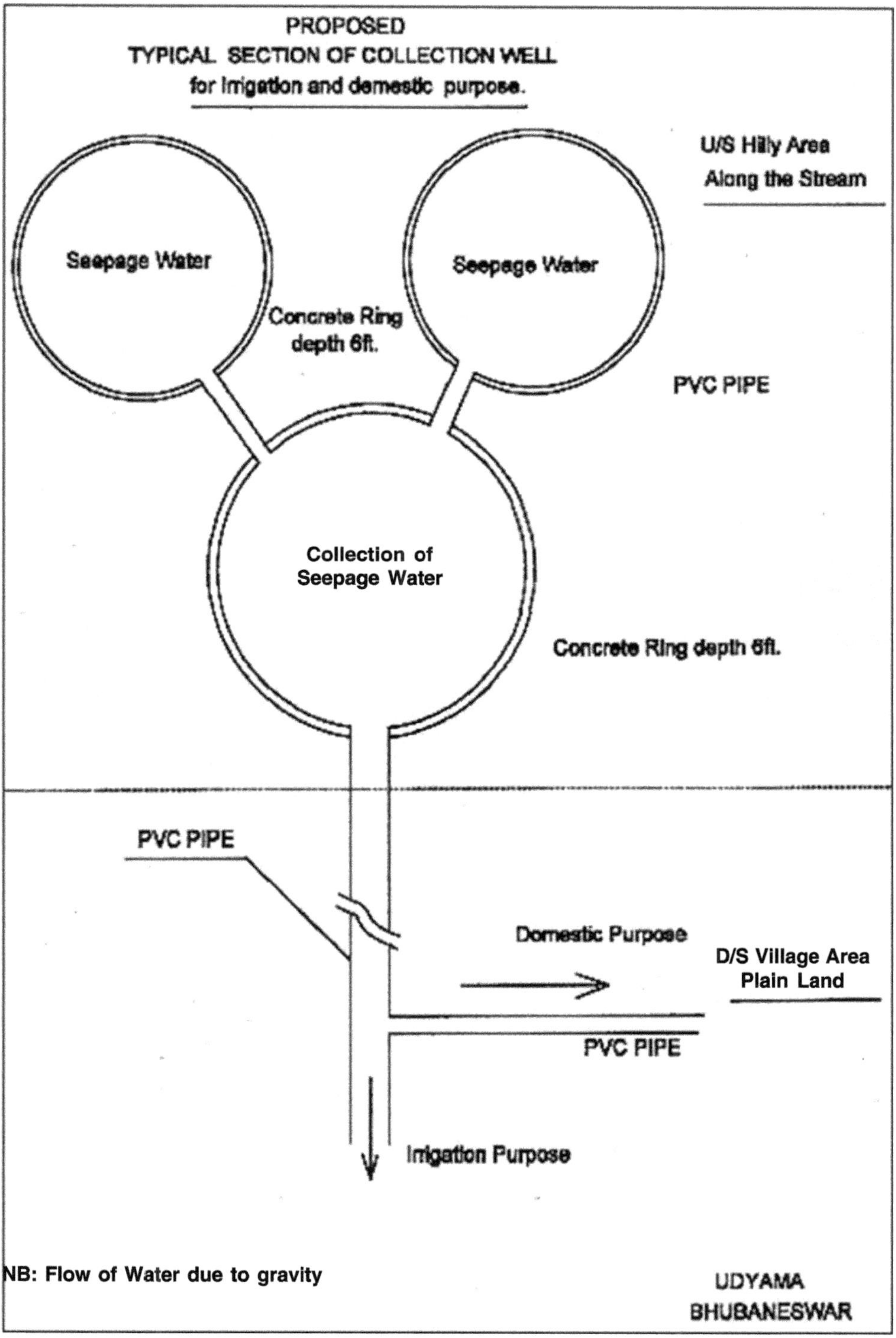
PROPOSED
TYPICAL SECTION OF COLLECTION WELL
for Irrigation and domestic purpose.
U/S Hilly Area
Along the Stream
Seepage Water
Seepage Water
Concrete Ring
depth 6ft.
PVC PIPE
Collection of
Seepage Water
Concrete Ring depth 6ft.
PVC PIPE
Domestic Purpose
D/S Village Area
Plain Land
PVC PIPE
Irrigation Purpose
NB: Flow of Water due to gravity
UDYAMA
BHUBANESWAR

Figure 5.3

Figure 5.4: Sustainable Model of Village Based DBI Convergence with Agri-Horticultural System.

8. Total sample size taken for study: 10 per cent from the total beneficiary (*i.e.* 2180 × 10/100=218 approx for effective study).
9. Case study added for ornaments of research.

Results and Discussion

By conducting above study in live, the authors were able to find out the following major findings for conducting this type of study in Hilly areas to provide low cost technology Irrigation to the people. This is the Social Innovation made by the Leadership of Young Scientist Bibhu *et al.* (2013-14) in Nayagarh Dist. of Odisha.

Personal Socio-Economic Profile of People with this Project

1. 78 per cent of people were belonging to Schedule Tribe followed by 13 per cent Schedule caste and rest 9 per cent belong to Other Castes. This is because of People's were aboriginals and belong to hilly forest areas.
2. Living status was very poor than the Open Category peoples, majority (98.7 per cent) people were belong to poverty line due to Illiteracy and Localite in nature and residing in Remote Pocket of the Dist. So Govt. is unable to access them.
3. 95 per cent people believe in their culture, tradition and taboos having Tribal Cottage and Mud houses. Only 5 per cent people were having Pucca House due to grant given by Govt. after Phailine Devastation under Indira Awas Yojana.
4. By making trend analysis we were able to find out the details of Forest, their livelihood, first matriculate, first technology adoption and seasonality of Springs and water resources there.
5. By using Problematic Tree Analysis we were getting the need to be addressed before Govt. and Policy makers about Primary health Centre, School, Electricity, Connectivity and upliftment of all basic needs they had deserved.
6. By using PRA analysis we had found that they were still using their traditional variety of seed, domestic animals and traditional ITK for doing every activity and they also using Local Medicine called as "Desi vaidya Aushadh".

How Community based Micro Diversion Based Irrigation made Positive Impact on People and Shows the Path of Success?

As Research is a live research conducted in Tribal Pockets, first the researchers were making awareness programme on various social aspects then after focused group discussion the research was implemented in said areas by Peoples Participation mode being convergence with Udyama a Local NGO being funded by SDTT Trust. The Technical help and advice was given by Researcher in every aspect. Then the Irrigation structures were made. Among all 60 per cent above structures were functioning well and able to distribute the Irrigation water for farming and local use. Experiment had been taken to supply the Hygienic water for drinking purpose. In this flow the people were strengthened by forming Water User Groups and Farmer's Clubs and defunct S.H.G.s were uplifted. Regular meeting and demonstration had been increased the people's interest, trust and participation. Now 2180 people were benefitted by this structure by doing two times cultivation of cereal crops along with diversification of vegetable farming by implementing Nutri- Farming and Kitchen garden in their field and also making waste land development by adding shelter belts to minimize the Disaster. 98 perc ent were practising organic way of cultivation along with traditional storage practices. Only 35 per cent people were able to grow their live status still today and 99.5 per cent were become self sufficiency to take their food safely to maintain food security. 20 per cent people were able to send their child to nearest Ashram school and Anganwadi for their education. After celebrating Vanomahostav, International Water Day, International Hand wash day they were become aware about nature, environment and climate for the first time with valid explanation. Now they were in a stage to uplift themselves with social prestige by maintaining their savings, records and fight for their rights.This Project would create a ray of hope for all beneficiaries. Now Govt. and NGOs were stretching their hands with funding for taking the responsibility of development in these areas after a Local Seminar being organized by the Leadership of Young Scientist Bibhu *et al.* with the help of Udyama, SDTT in particular and Kyobo foundation with UNEP-10 YFP in general in awareness program.

Concluding Remarks

Here the Key Researcher want to submit one thing a technology should be socially acceptable, technically feasible, economically viable and climatically compatible. In each project the Project should be framed by People-Led approach in order to achieve Sustainable Development Goals hence be justified. So in Hilly and poverty stricken areas this Low Coast Community centred Micro Diversion based Irrigation system must act as the only Panacea for supplying water and uplifting the people.

Bridge The Gap.......Fill the Gap Concept being invented by Villagers of Kandhakhalamada as a Case Study

- ☆ Kandhakhalamada, a typical Tribal Village in Nayagarh District mostly dominated by Kandha Tribes. Though it is a village in District Map but still it has zero existence in terms of Development. They are living with their own way and own policy. Govt. Policies are still now unreachable, inaccessible to this area. Improper communication facilities to this area imparts lower human develop index (HDI). So still after 66 years independence, they are remain as such in dark era. The village is surrounded by green forest and the road to this village from main road is very pathetic, dangerous and out of order during rainy season.So this road is always remain detached from the world many times. This is the main cause of their undevelopment, negligence from all. But most of the People are depending on this Path to do their necessary fulfillments.There is a saying "United we stand, divided we fall". This concept was sudden came to the mind of a Tribal Leader named Bhramara Malik. So he had shared this topic and discussed with all villagers of his village at a common meeting. Finally, they had decided to contribute their labour on rotation basis to complete the task of Local road preparation by their own way means self-help mode without any aid from Govt./NGO/Other sources. Both Male and Female were involving in this activity for 5 days. They had collected the Pebbles, sand stones, Sand, big rocks and spread it over the narrow foot path to become it wide. They were successful by intervening this concept "Bridge the gap.........Fill the gap" by self help mode. Now they were very happy to moving on their own developed way with personal satisfaction and pride without begging any help from other. This is the best example for other inaccessible area to prepare their own way own life by this concept.

- ✰ Now it is the time to congratulate the mentors of this mission. They were Basanti Jani, a tribal women farmer, Binod Malik, Bhramara Malik and all credits goes to "Kandhakhalamada" villagers.
- ✰ But Bibhu Santosh Behera, a leading Young Scientist of OUAT, Bhubaneswar put first step to file this case study for drawing attention of Govt./any other sources for bridging their gap..........filling their gap by stretching hand to this area by making a Concrete Road and Bridges on local water basin for which the people can able to access outside and develop accordingly. This concept is also in the line of "For the people, Of the People and By the People" hence be justified.

Acknowledgement

I acknowledged the help of My University, my teachers and help of Co-authors in general and Peoples of this area in particular. I am thankful to Mr. Pradeep Mohapatra, Secretary Udyama for his all type help. Finally, I bow my head before almighty to make this mission a grand success. This is dedicated to all innocent people.

References

1. Behera, B.S. (2013). "Correlates of Adoption of tribal Vegetable Farmers of Keonjhar District of Odisha" (Thesis of OUAT, Bhubaneswar).
2. Locating the Poor: Spatially Disaggregated Poverty Maps for Sri Lanka. Upali A. Amarasinghe, Madar Samad and Markandu Anputhas. 2006.
3. Strategies to Mitigate Secondary Salinization in the Indus Basin of Pakistan: A Selective Review. M. Aslam and S. A. Prathapar. 2006.
4. Multiple-Use Water Services to Advance the Millennium Development Goals. Barbara van Koppen, Patrick Moriarty and Eline Boelee. 2006.

5. Irrigation and Schistosomiasis in Africa: Ecological Aspects. Eline Boelee and Henry Madsen. 2006.
6. The Reliability Improvement in irrigation Services: Application of Rotational Water Distribution to Tertiary Canals in Central Asia. Iskandar Abdullaev, Mehmood U I. Hassan, Herath Manthrithilake and Murat Yakubov. 2006.
7. Carbon, Land and Water: A Gobal Analysis of the Hydrologic Dimensions of Climate Change Mitigation through Afforestation/Reforestation. Robert J. Zomer, Antonio Trabucco, Oliver van Straaten and Deborah A. Bossio. 2006.
8. Informal Irrigation in Urban West Africa: An Overview. Pay Drechsel, Sophie Graefe, Moise Sonou and Olufunke O. Cofie. 2006.
9. Malaria Mosquito Resistance to Agricultural Insecticides: Risk Area Mapping in Thailand. Hans. J. Overgaard. 2006.
10. Use of a Hydrological Model for Environmental Management of the Usangu Wetlands, Tanzania, Japhet J. Kashaigili, Matthew P. McCartney, Henry F. Mahoo, Bruce A. Lankford, Boniface P. Mbilinyi, Daniel K. Yawson and Siza D. Tumbo. 2006.
11. An Irrigated Area Map of the World (1999) Derived from Remote Sensing. Thenkabail, P. S., Biradar, C. M., Turral, H., Noojipady, P., Li, Y. J., Vithanage, J., Dheeravath, V., Velpuri, M., Schull, M., Cai, X. L., and Dutta, R. 2006.
12. Behera, B.S., Behera,A.C., Behera, R.A. (2011) Irrigation Strategies for Dryland areas in Orissa as a review(Abstract) Published in OEC Conference Proceedings.

Web and Other Sources

1. www.baif.org
2. www.livolinkfoundation.com
3. www.sdtt.org
4. www.cera.com
5. www.jgateplus.com
6. www.cera.iari.res.in
7. www.agropedia.iitk.ac.in
8. www.agricola.nal.usda.gov
9. http://www. odi.org.uk/resources/download/164.pdf
10. http://www.itu.int/wsis/docs/geneva/official/poa.html
11. http://www.itu.int/wsis/docs2/tunis/off/6rev1.html
12. http://agropedia.iitk.ac.in/openaccess/sites/default/files/WS per cent 205.pdf.
13. http://www.communityict.ca/docs/daknet-case.pdf.

Chapter 6

Repercussions of Mobile Phone Addiction among Teenagers: Special Study among Students of Selected Schools of Indore City of Madhya Pradesh

Neeta Deshmukh

Department of Home Science,
DAVV-Sugnidevi Girls College, Indore, M.P.

The use of mobile phone is increasing day by day. Today mobile phone is one the major necessity of life. But in spite of all its benefits and plus points mobile phone use has very dangerous and destructive effects on our society. It has totally disturbed the structure and foundation of our society. The major and most destructive effects of mobile phone use on our society are on teenagers. Mobile phones are not just a substitute of the traditional red phone box seen on the streets. The mobile phone has become a life support for many people. People of all ages carry handsets for all different reasons: business people may use it to plan day to day appointments and meetings, Pleasure users may use it for tools for just emergencies, teenagers may use it to text friends, listen to music or take pictures, whilst children may just use it to play games. There are many advantages and disadvantages of owning a mobile phone.

Students do not give proper time to their studies and waste their time in playing games, listening music, watching videos and reading messages on their mobile phones. Parents are not aware about the activities of their children. Teenagers' boys and girls communicate each other through their mobile phones and do negative kind of activities. This is very dangerous and major negative impact of mobile phone on our society. Our new generation is totally destroyed by this negative aspect of mobile phones. Mobile phones now becoming a major communication tool for criminal's activities. People are using this fastest communication for terrorism and other kind of criminal activities. We must overcome these negative impacts of mobile phone use on our society otherwise our society and new generations will completely destroy.

The students are just addicted to mobile phones. They can be seen playing games, chatting, and talking to their friends on their mobile phones most of the times. This is the reason why they don't get time for studies. In fact, students are more interested in wasting their time on mobile phones, rather than spending it on studying.

Materials and Methods

The sole objective of this study is to assecs the impact of the use of mobile phone among teenagers especially those studying in from class 8 to 12 and their age varied from 13 to 18 years. In this study they were 200 students from 8 to 12 standards those selected from different two convent and government school of Indore city of Madhya Pradesh.

To assess the significant difference among them having regular and excessive use of mobile phone rangers from normal to excessive night use, the data of study were selected through multi variate random sampling method and questionnaire to find out impart of variables. The selection criteria had been to find out the significant difference of variable depression, isomnia irretation and health impact.

To assess the difference the study took 3 point rating scale method and t-Test to conclude specific differentiation.

Observation

During the study following observations have been found to mark specific.

1. It is observed that mean age of experimental and controlled group were 13 and 18 years respectively. All respondents were students.
2. It is also founds that the respondents those used excessive mobile phone regularly compared to who used less, through combined experimental group with 9.1287093 to 6.454972 respectively.
3. The study revealed that mean value of Day using was 125 and A2 mean value was found 8.75 standard deviation were found 6.454972 and 4.787136 for A1 and A2 respectively. While T-value of A1 and A2 were found 0.933257.
4. The study evidently found that frequency of contents sharing of with entertainment and adult and study content have significant difference. It has been found that the respondents those shared and used entertainment adult contents had revealed significant difference compared to chose shared study contents. The mean varied from 22.5 to 27.5 of A1 and A2 respectively with 11.90238 and 11.90238 SD the t-value of contract sharing was found 0.59409.

5. The study evidently found the impact of psychological and physical health of mobile using teenagers. Depression, Insomnia and irritation had been observed from mean values of 17.5, 20 and 12.5 for A1, A2 and A3 respondents respectively. While the study found 6.454972, 10.80123 and 6-454972 for A1, A2 and A3 respectively. The t-Value was found 0.397636 for level of depression and insomnia while the significant difference was found in depression and irritation level with has from of 175 and 12.5 respectively while the level of t-value was 1.090445.

Frequency Table of Mobile Using

Sl.No.	*Respondents*	*Day Users*		*Night Users*		
		A-1	*A-2*	*A-3*	*A-4*	*Total*
1.	B(CS)	5	10	10	25	50
2.	B(GS)	15	5	25	5	50
3.	G(CS)	10	15	5	20	50
4.	G(GS)	20	5	15	10	50
5.	TOTAL	50	35	50	60	300
6.	STD DEV	6.454972	4.787136	6.454972	9.1287093	

T-Value of A1 Vs A2

Mean of A1	12.5
Mean of A2	8.75
Std Dev of A1	6.454972
Std Dev of A2	4.787136
S1=S(A1)*S(A1)	41.66667
S2=S(A2)*S(A2)	22.91667
D1=S1/4	10.41667
D2=S2/4	5.729167
H=SQR(D1+D2)	4.018188

T-Value of A2 Vs A3

Mean of A2	8.75
Mean of A3	12.5
Std Dev of A2	4.787136
Std Dev of A3	6.454972
S1=S (A2)*S (A2)	22.91667
S2=S (A3)*S (A3)	41.66667
D1=S1/4	5.729167
D2=S2/4	10.41667
H=SQR(D1+D2)	4.018188
T-Value	–0.93326

T-Value of A1 Vs A3

Mean of A1	12.5
Mean of A3	12.5
Std Dev of A1	6.45497224
Std Dev of A3	6.45497224
S1=S(A1)*S(A1)	41.6666667
S2=S(A3)*S(A3)	41.6666667
Da=S1/4	10.4166667
Ds=S2/4	10.4166667
H=SQR(D1+D2)	4.56435465
T-Value	0

T-Value of A2 Vs A4

Mean of A2	8.75
Mean of A4	15
Std Dev of A2	4.787136
Std Dev of A4	9.128709
S1=S (A2)*S (A2)	22.91667
S2=S (A4)*S (A4)	83.33333
D1=S1/4	5.729167
D2=S2/4	20.83333
H=SQR(D1+D2)	5.153882
T-Value	–1.21268

T-Value of A1 Vs A4

Mean of A1	12.5
Mean of A4	15
Std Dev of A1	6.454972
Std Dev of A4	9.128709
S1=S (A1)*S (A1)	41.66667
S2=S (A4)*S (A4)	83.33333
D1=S1/4	10.41667
D2=S2/4	20.83333
H=SQR(D1+D2)	5.59017
T-Value	–0.44721

T-Value of A3 Vs A4

Mean of A3	12.5
Mean of A4	15
Std Dev of A3	6.454972
Std Dev of A4	9.128709
S1=S (A3)*S (A3)	41.66667
S2=S (A4)*S (A4)	83.33333
D1=S1/4	10.41667
D2=S2/4	20.83333
H=SQR(D1+D2)	5.59017
T-Value	–0.44721

Frequency Table of Content Sharing

Sl.No.	*Entertainment and Adult*	*Study*		
		A-1	*A-2*	*Total*
1.	B(CS)	35	15	50
2.	B(GS)	15	35	50
3.	G(CS)	30	20	50
4.	G(GS)	10	40	50
5.	TOTAL	90	110	200
6.	STD DEV	11.90238	11.90238	

T-Value of A1 Vs A2

Mean of A1	22.5
Mean of A2	27.5
Dev of A1	11.90238
Std Dev of A2	11.90238
S1=S (A1)*S (A1)	141.6667
S2=S (A2)*S (A2)	141.6667
D1=S1/4	35.41667
D2=S2/4	35.41667
H=SQR(D1+D2)	8.416254
T-Value	–0.59409

Discussion

The regular using of mobile phone among teenagers has been much common nowadays and specially students. Excessive user of mobile phones has had negative impact on psychological and physical health of serious health user such as depression, insomnia, radiation. There has been significant difference among respondent of these study. The impact of excessive leading of mobile phones caused a lot of health since from depression, insomnia. The content sharing of entertainment adult drove respondents of excessive the component of study contents of less user. The significant difference had also been observed that depression plays vital role in teenagers.

Impact of Psychological and Physical Health

Sl.No.	*Respondents*	*Depression*	*Insomnia*	*Irritation*	
		A-1	*A-2*	*A-3*	*Total*
1.	B(CS)	10	35	5	50
2.	B(GS)	25	15	10	50
3.	G(CS)	15	20	15	50
4.	G(GS)	20	10	20	50
5.	TOTAL	70	80	50	200
6.	STD DEV	6.454972	10.80123	6.454972	

T-Value of A1 Vs A2

Mean of A1	17.5
Mean of A2	20
Std Dev of A1	6.454972
Std Dev of A2	10.80123
S1=S (A1)*S (A1)	41.66667
S2=S (A2)*S (A2)	116.6667
D1=S1/4	10.41667
D2=S2/4	29.16667
H=SQR(D1+D2)	6.291529
T-Value	–0.39736

T-Value of A2 Vs A3

Mean of A2	20
Mean of A3	12.5
Std Dev of A2	10.80123
Std Dev of A3	6.454972
S1=S (A2)*S (A2)	116.6667
S2=S (A3)*S (A3)	41.66667
D1=S2/4	29.16667
D2=S3/4	10.41667
H=SQR(D1+D2)	6.291529
T-Value	1.192079

T-Value of A1 Vs A3

Mean of A1	17.5
Mean of A3	12.5
Std Dev of A1	6.454972
Std Dev of A3	6.454972
S1=S (A1)*S (A1)	41.66667
S2=S (A3)*S (A3)	41.66667
D1=S1/4	10.41667
D2=S3/4	10.41667
H=SQR(D1+D3)	4.564355
T-Value	1.095445

Conclusion

The study indicates that a great impact on teenager chose enjoyed excessive using of mobile phones including day and night hours. It has also been noted that those using mobile phones using entertainment and adult contents sharing have been found more choke to psychological and physical health users. Compared to those used study contents. Entertainment contents that include games, video clips of song and porn movies caused negative impact and led them to significant intolerance and depression levels.

The addiction of excessive using of mobile phones and sharing entertainment contents among teenagers caused them insomnia compare to those who were study contents is limited hours. If teenager have to divert some different indoor and outdoor activities for few hours then chat could be effective way to keep at the bay of the negative impact of mobile phone.

References

1. American Academy of Sleep Medicine. "Excessive Mobile Phone Use Affects Sleep In Teens, Study Finds." Science Daily. Science Daily, 9 June 2008. www.sciencedaily.com/releases/2008/06/080609071402.htm
2. http://www.sciencedaily.com/releases/2008/06/080609071402.htm
3. "Conclusions on mobile phones and radio frequency fields". European Commission Scientific Committee on Emerging and Newly Identified Health Risks (SCENIHR). Retrieved 8 December 2008.
4. International Commission on Non-Ionizing Radiation Protection (April 1998). "Guidelines For Limiting Exposure To Time-Varying Electric, Magnetic, And Electromagnetic Fields (up to 300 GHz)" (PDF). *Health Physics* **74** (4): 494–505.PMID 9525427. Retrieved 28 March 2010.
5. Foster, Kenneth R.; Repacholi, Michael H. (2004). "Biological Effects of Radiofrequency Fields: Does Modulation Matter?".*Radiation Research* **162** (2): 219–25. doi:10.1667/RR3191. PMID 15387150.

6. Glaser, Roland (December 2005). "Are thermoreceptors responsible for "non-thermal" effects of RF fields?" (PDF). *Edition Wissenschaft* (Bonn, Germany: Forschungsgemeinschaft Funk) (21). OCLC 179908725. Retrieved 19 January 2008.

7. Blank, Martin; Goodman, Reba (2009). "Electromagnetic fields stress living cells". *Pathophysiology* **16** (2–3): 71–8.doi:10.1016/j.pathophys.2009.01.006. PMID 19268550.

Chapter 7

Stored Grain Insect Pest and their Effective Ecofriendly Management

Bibhu Santosh Behera[1], Soubhagya Behera[2] and Subhashree Dash[3]

[1]Department of Extension Education, College of Agriculture, OUAT, Bhubaneswar – 751 003, Odisha
[2]AAO, Thuamul Rampur, Odisha
[3]Department of Entomology, OUAT, Bhubaneswar – 751 003, Odisha

ABSTRACT

Ecofriendly management of stored-grain pest can be done by holistic approach by using less residual insecticide or pesticide as per the incidence of insect-pest with Economic threshold level and Economic injury level by adopting bio control measures and cultural Indigenous technical knowledge. Here researchers want to narrate the effective ecofriendly management practices to help farmer friends.

***Key words**: Ecofriendly approaches, Stored grain pest, ITK.*

Introduction

India is basically an agricultural country and it has most variable climatic regions owing to its geographic features. Total arable land area is 168 million hectares and major part of it falling under tropical climate, and a variety of cereals, oil seeds, pulses, vegetable and horticultural crops are being cultivated. India has achieved self sufficiency in food grains but there is an urgent need to improve your productivity in all crops to meet future challenges. Maintenance of good food grain quality during long term storage has become one of the major considerations in food security and safety planning. Failure to comply with good storage management practice can jeopardize food supply chain. The many difference types of insects that invade stored dried foods are often referred to as "pantry pests". They contaminate more food then they consume, and most people find the contaminated products unfit for consumption. Most stored-grain insect infestations originate in the immediate area of the storage facilities; therefore, area sanitation is important. Many of the common stored-grain pests have the ability to move from one facility to another. Practices that limit the pests access to food and shelter will help reduce the potential for future infestations. Food product that are left undisturbed on the shelves for long periods are particularly susceptible to infestations. However, foods of any type can become infested. Managing stored grains requires the use of various techniques to ensure that the quality of the grain entering the storage facilities does not deteriorate over time. These measures include: the use of sanitation; storing sound dry grain, managing temperature and aerations, and using chemical Protestants, regular sampling and fumigation. Storage bin facilities play an important role in determining whether grain quality is maintained and should be inspected regularly. Regular monitoring will help to ensure that grain quality will be maintained at the highest level possible. It is estimated that 60-70 per cent of total food grains produced in the country is stored at farmer's level in traditional old receptacles. According to FAO estimated the world wild annual losses in storage is around 10 per cent in all stored grains. About 13 millions tones is damage only by insect and 100 million tons by storage. In realising this, one of the importance stumbling blocks seems to be the stored grain losses due to insect pests. There is an urgent need to assess such losses in order to frame strategies to overcome then.

Methodology

Here the secondary sources of information is being used to prescribe the suitable effective ecofriendly strategies for farmers.

Source of Infestation

There are three sources of infestation are elaborated below:

1. In the field condition (Rice weevils, pulse beetle and grain moth)
2. Infestation in godowns from the pre existing pests and
3. Infestation from a storage bags and containers.
4. Some insets like rice weevil, grain moth and pulse beetle starts laying eggs on the maturing grains in the field itself and are thus carried to godown.

Type of Grain Insects

They are two types as cited below:

1. *Primary grain insects*: It refers to a group of insects that attack whole undamaged grains *i.e.* Rice weevils, bean weevils and lesser grain borers.
2. *Secondary grain insects*: It refers to a group or complex of insects that feed on fragments of grains and cereals *i.e.* grain moths, flour beetles and Indian meal moth.

Apart from these the insects are also classified as minor and major types.

1. *Minor insects*: for example:- Rice moth *Corcyra cephalonica* (Lepidoptera)
2. *Major insects*: for example:-
   a. Indian meal moth (*Plodia interpunctella*) Lepidoptera
   b. Red rust flour beetle (*Tribolium castaneum*) (coleoptera)
   c. Rice weevils (*Sitophilus oryzae*) (Coleoptera)
   d. Angoumois grain moth (*Sitotroga cerealella*) (Lepidoptera)
   e. Khapra beetle (*Trogoderma granarium*) (Coleoptera)
   f. Lesser grain borer (*Rhyzopertha dominica*) (Coleoptera)
   g. Pulse beetle or gram/mung dhora (*Callosobruchus chinensis*) Coleoptera

Types of Grain Losses

There are two types of grain losses:

1. Quantitative Loss

- ☆ Loss weight due to reduction in moisture content of the grain.
- ☆ Losses due to grain depredation by insect, rodents and birds.
- ☆ Losses due to grain mis-handling.

2. Qualitative Loss

- ☆ Discolouration of grains due to development of mould or faulty method of grain drying.
- ☆ Loss due to deterioration in nutritional quality.
- ☆ Loss due to presence of insect grain fragments, excreta, urine etc. A few beetles carry toxins on their body, which they are capable of imparting to the infested grains.
- ☆ Losses due to presence of aflatoxin produced by micro-organisms like fungus, *Aspergillus* sp.

Factor Affecting Infestation

Like all living organisms, Insects have 3 basic requirements besides food to be able to stay alive: water, oxygen and temperature. Moisture is obtained from two sources: Atmosphere and metabolism while the atmospheric moistures is absorbed by the containers and passed on to the grains hence to the insect pest who feed on them.

Grain Moisture

As per Agricultural thumb rule, grain stored for extended periods of time should contains no more than 12 per cent moisture. Insects and fungi do not develop well in grain with moisture content of 12 per cent or below. The minimum moisture content required by some storage pest are given in tabular form below.

Table 6.1

Sl.No.	Insect Name	Moisture Per cent
1.	*Trogoderma granarium*	0-1.9
2.	*Sitophilus oryzae*	9.5-11
3.	*Rhyzopertha dominica*	9.0-10
4.	*Cadra cautella*	10.0
5.	*Tribolium castanium*	10.0
6.	*Corcyra cephaonica*	9.0

Table 6.2: Upper Limit of Grain Moisture Content for Safe Storage

Commodity	Moisture Content (Per cent wet basis)
Paddy, rice(raw)	14
Rice(parboiled)	15
Wheat, Bengal gram, kabuligram	12
Sorghum, maize, barley, ragi, bajra, pulse, turmeric, wheat atta maida besan	12.5
Coriander, chillies	10
Groundnut pods	6-7
Mustard seed	5-6

Temperature and Aeration

Temperature is the most important factor for controlling the development and multiplications. Grains harvested and stored in the hottest part of the year stand a greater chance of becoming infested, since insects reproduce rapidly at temperatures in the range of 60°-90°F. The use of an aeration controller is extremely helpful when managing aeration for insect control. Controllers start aeration fans when outside temperatures reach a preset point. Small reductions (10°-15°F) in grain temperature have proven to significantly reduce the reproductive rates of stored-grain insects and reduce insect damage. Several practices should be followed when filling bins to permit even aeration of the grain mass. Therefore, efforts should be made to prevent the heat spots by way of cooling which may be done by forced aeration.

Art of Prevention and Management of Stored Grain Pests

Sanitation

Before adding grains in to storage facilities, make sure it is clean and free from insects, old grain, trash and be sure the floors, ceiling, windows, ledges, walls, loading/unloading equipments are clean. The area outside the storage room or bin should also be free of weeds, insects and others. Insects can breed and persist in these areas and infest new grain when it's placed into the storage bins. It is best to clean and treat storage facilities at least two weeks prior to adding new grain.

Preventive Measures

1. Godowns should be cleaned.
2. All the cracks crevices and holes should be filled with mud or cement.
3. The godown should be spread the insecticides before storage of food grain
   - ☆ The ceiling and walls of the godown may be sprayed with 0.5 per cent malathion @ 3 l/m^2.
   - ☆ Fumigation with EDCT mixture for 24 hours @ 10 litre/40cubic meter of space.
4. Cleaning of bags before storage
   - ☆ By dipping them in boiling water for 15-20 minutes.
   - ☆ Drying then in hot sun heat for 6 hours
   - ☆ By dipping them in to 1 per cent malathion solution for 10 minute
5. Bin facilities should be weather tight, rodent proof, steel, and on moisture-proof concrete base. Bins should be equipped with a perforated-floor aeration system and weather-proof roof vents. All bins should be inspected on a regular basis to guard against leaks, condensation, and deterioration of any kind. Once filled, attempt to seal the bottom and sides of the bin so insects and rodents can only enter the top of the facility. Do not seal roof aeration exhaust or inlet vents except during fumigation so the top of the bin can be easily sampled and top dressings applied if necessary.
6 Empty-bin Treatment: insecticide applications are encouraged to prevent the infestation Empty-bin sprays are highly recommended when grain is stored in summer, if there are areas difficult to clean, or if there has been a history of insect problems. After bins have been properly cleaned and inspected and prior to adding new grain, treat the facility with a labelled insecticide. Spray to run-off the inside surface and as much of the outside, including the nearby ground surfaces, aeration ducts, and grain handling equipment, as possible. Sprays should be concentrated on cracks, crevices, and areas difficult to clean. Applications should be made at least two weeks prior to adding new grain. Allow 24 hours for sprays to dry. These treatments provide a barrier to insects that may be attracted to the storage facilities and also provide control of the insects not removed during the cleaning operation.

Remedial Measure

When the infestation of stored insects is in progress used the following insecticides

1. Aluminium phosphide tablets @ 1/metric ton of grain or 7 tablets/28 m^3
2. EDB ample @ 1/q of grain.

Fumigation

Fumigation should only be conducted by the trained, experienced and registered applicators.

Methyl Bromide Fumigants

Under most conditions, Fumigation with methyl bromide will not harm germination. However, high doses (generally used for insects) for more than 24 hours coupled with temperatures above 85°F and moisture greater than 12 per cent can negatively impact seed germination. The amount of fumigant required is determined based on the size of the godown and environmental conditions present. The cylinder is placed on a set of scales and the weight determined. The valve is opened and then closed once the proper amount of the fumigant has been released.

Phosphine Fumigants

Phosphine has no adverse effects on seed germination when applied according to label directions and at labelled rates. Since the gas diffuses through the grain rapidly, structures must be sealed properly, especially under cooler conditions.

Ecofriendly Approaches

By adopting the Biocontrol and Cultural measures impregnated of ITK in that theme by adopting Tricho cards, Pheromone traps, Colour traps and Light traps the pest can crops can be controlled. It has less biomagnifications and ecohazards and can use as sustainable measures. The ITK may be used here is the fumigation of Chilli, and narcotics.

Conclusion

India produces about 210 million tones of food grains per year. Production has been steadily increasing due to advancement in production technology, but losses have remained static at 10 per cent. This means that the loss of food grains is also increasing with the increasing in food production. The main reason for this is improper storage, and an average of 6 per cent out of a total 10 per cent loss takes place during storage of food grains. Large-scale storage is an essential component in both food safety and security systems. It can also enhance grain marketing chains. However, maintenance of stored grain quality under large-scale storage system can pose big risk unless measures are taken to address deteriorative changes during storage. The use of aeration is an effective and cheap technique for maintaining grain quality during large-scale and long-term bulk storage of grains. However, it was not found to be effective against insect growth. Insect control is therefore needed to be integrated in any aeration practices. However, the insect activity was not found to be detrimental to the final quality in the first six months of storage period.

References

1. Katyayan, A. Fundamentals of Agriculture Volume II.
2. Tiwari, G., Shukla, D., and Verma, G.K Stored insect pest and management, Kurukshetra August 2010 publication.
3. Shrivastav, K.P Applied Entomology Volume II.
4. Other sources from–Agrobios and Agriculture today magazines and from Cera web mail.

Chapter 8

Determination of Water Quality of Surface Water in Eight Blocks of Banda Region: Chemical and Biological Studies

Chhavi Purwar and Preety Agarwal

Department of Chemistry, Pt. J.N.P.G. College,
Banda – 210 001, U.P.
Department of Teacher Education,
D.A.V.P.G.College, Muzaffarnagar, U.P.

ABSTRACT

The investigation was undertaken with an objective of developing techniques for the potability of drinking water of eight blocks of Banda region. For this investigation the water samples of these blocks have been collected on Sept-2012. In chemical analysis, the pH value, turbidity, temperature, alkalinity, total hardness, total dissolved solid (TDS), total suspended solids (TSS), total solid presence, conductivity, chloride, fluoride, nitrite, nitrate, calcium, magnesium, iron content etc. have been determined. In bacteriological analysis, bascilli, navicula, anavina, coccus have been determined by using presumptive test, confirmed coliform test and completed coliform test. It was found from the chemical analysis that surface water is good for drinking water.

Keywords: *Potability, Surface water, Banda city.*

Introduction

Water is essential for survival of any form of life. Water present in lakes and ponds is called still water. Water present in rivers and streams is called running water, while water present in rains, snow, dew etc. is called fresh water. Water present in all these water bodies is called surface water. It lays the largest role of shaping the geography of land.[1] Water is a universal solvent and renewable resource. Any unusual activities of man, which make water unfit for all living beings, directly or indirectly cause water pollution.

Due to increasing urbanization, cultivation and industrialization have caused enormous microbiological and chemical pollution of surface water. Microbiological contamination causes variety of water borne diseases. Certain microorganisms present in water, especially pathogenic species, cause diseases of man and animals and can be referred to as bio-pollutants.

Addition of any unwanted substance which changes the composition of water, smell, taste, colour, pH or dissolved oxygen cause water pollution. Reduction of oxygen, addition of pathogens and increase of toxicity in water which may alter the physical or biological properties of water are responsible for water pollution.[2]

Results and Discussion

The potability of water demands that it must be more or less free from turbidity, colour, taste and odour, well aerated and it must be free from pathogenic microorganisms and harmful chemicals. The eight samples were collected from eight blocks of Banda region on September -2012.

Water Quality Parameters

Following water quality parameters were analysed.

Temperature

Temperature is basically important for the chemical and biological reactions of organism in water. The increase in temperature decreases the potability of water because of elevated temperature, carbon dioxide and other volatile gases which impart taste are expelled. The maximum temperature of surface water in study areas was found as 26 °C in the month of September-2012.

Colour

In all types of water samples no colour is found. Colour in drinking water may be due to the presence of coloured organic substance which is not related to health. So it can be concluded that all these sources of drinking water are safe with respect to colour. Few water samples are muddy which can be clear through filtration.

Total Solid, Total Dissolved Solid and Total Suspended Solid

TDS values indicate the general nature of water quality and are usually related to conductivity. The experimental result of total solid concentrations, dissolved solid concentrations and suspended solid concentrations are given in Table 8.1. This table indicate that the amount of suspended substances are higher than the dissolved substances. Among different kinds of solid mainly total dissolved solid present in the drinking water is more important. Water with high dissolved solids has inferior potability and may reduce the unfavourable physiological reaction in the transient

Table 8.1: Chemical Analysis of Surface Water Collected from Eight Blocks of Banda Region on September 2012

Parameters	S_1	S_2	S_3	S_4	S_5	S_6	S_7	S_8
Sample name	Jaspura	Tindwari	Baberu	Kamasin	Mahua	Visanda	Badokhar Khurd	Naraini
Temperature °C	25.2	26.1	26.1	25.8	25.9	25.8	26	26.1
Colour	Muddy	Colourless	Colourless	Muddy	Muddy	Muddy	Colourless	Colourless
Total solid (mg/L)	305	450	320	412	454	320	300	290
Total dissolved solid (mg/L)	105	150	120	112	154	120	100	90
Total suspended solid (mg/L)	200	300	200	300	300	200	200	200
pH	8.01	6.89	8.03	8.15	7.68	7.6	8.5	7.1
Conductivity (mhos)	0.6	0.29	0.24	0.23	0.29	0.25	0.27	1.2
Total hardness (mg/L)	400	200	400	220	300	600	200	400
Ca -hardness (mg/L)	200	100	200	110	150	300	100	200
Mg -hardness (mg/L)	200	100	200	110	150	300	100	200
Turbidity (NTU)	4.8	4	10.6	4.9	4	4.8	12	15
Akalinity (mg/L)	120	130	106	112	144	140	108	92
Chloride (mg/L)	20	50	20	20	30	30	40	34
Fluoride (mg/L)	0.4	0.4	0.4	0.4	0.4	0.4	0.4	0.4
Iron (mg/L)	0.6	0.8	0.6	0.6	0.8	0.6	0.8	0.5
Nitrite (mg/L)	0.2	NIL	NIL	0.1	0.5	NIL	NIL	NIL
Nitrate (mg/L)	10	NIL	NIL	12	13	NIL	NIL	NIL
Oxygen absorbed from $KMnO_4$	9.5	8.1	9.6	13.9	7.7	10	9	9.8

consumer[3]. Solids at higher levels create excessive hardness, poor taste, mineral deposition and corrosion may occur. It was also observed that solid substance present in sample in undissolved condition can be removed easily by simple filtration or sedimentation.

pH

pH is considered as an important ecological factor and provides an important piece factor and piece of information in many type of geochemical equilibrium or solubility calculation factor and piece. The water samples exhibited an alkaline pH in the range of 7.05 to 9.46. The sample of Tindwari (S_2) showing the pH 6.89. An acceptable range of pH of drinking water is from 6.5 to 8.5. Water with a pH below 6.5 is considered acidic and may cause heavy metal leaching[4] causing plumbing damage etc. In all water samples the pH lying in the limited range. No significant difference was noted in the observed pH, the variation in pH due to change in sampling location was also insignificant.

Conductivity

Electrical conductivity of water is a direct function of its total dissolved salt. Hence it is an index to represent the total concentration of soluble salt in water.The results of conductance in different water samples listed in Table 8.1. Conductivity measurements offer a rapid and nondestructive way to measure ion content in the sample. High conductance values also cause many diseases[5]. The conductance values are showing a limiting range in respect of presence of ions.

Total Hardness

The results of total hardness in water sample are shown in Table 8.1. The values of total hardness also indicate the heavy amount of dissolved substance in drinking

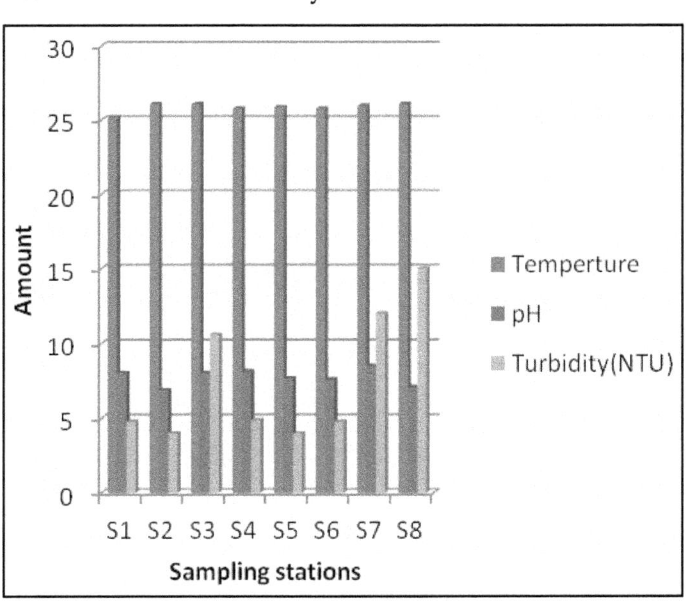

Figure 8.1

water. It is recommended that water from any source is not always fit for use unless treated for hardness[6]. The observed results also support the TDS result and conductivity result. The total hardness value also does not cross the permissible limit (Table 8.3).

Calcium Hardness

Calcium is one of the content found in surface water. The main source of calcium is dissolution of soil and rock. Due of presence of calcium incrustation in pipes and it becomes corrosive in combination with chloride and causes pitting of boilers. The maximum value of Ca as 300 mg/L was recorded in S_6 location and minimum 100 mg/L value was recorded in S_2 location.

Magnesium Hardness

Another content which is present in the water is magnesium. Dissolution of soil and rock are sources of magnesium. It becomes corrosive in combination with chloride and in combination with sulphate causes laxative effect. The maximum value of Mg as 300mg/L was recorded in S_6 location and minimum 100mg/L value was recorded in S_2 location.

Turbidity

Turbidity in water is caused by suspended matter like clay, silt, organic matter, phytoplanktons and other microscopic organisms. The maximum turbidity was found as 15NTU at location S_8 and minimum value was found as 4NTU at the location S_2 and S_5.

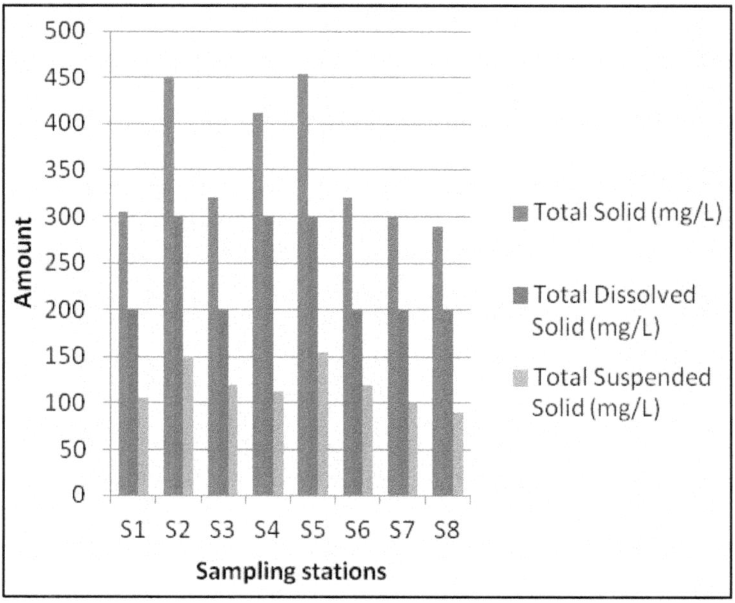

Figure 8.2

Alkalinity

Alkalinity value in water provide an idea of natural salts present in water. The cause of alkalinity is due to minerals which dissolve in water from soil. The various ionic species that contribute to alkalinity includes bicarbonate, hydroxide, phosphate, borate and organic acids. These factors are characteristics of the source of water and natural processes taking place at any given time. The maximum value of alkalinity of surface water was found as 144 mg/L for S_5 location and 92 mg/L for S_8 location.

Oxygen Absorbed form $KMnO_4$

It represents the amount of oxygen is dissolved in water. Clean surface water is generally found to be saturated with dissolved oxygen and have pleasant taste. Oxygen dissolves in water by absorption from air or by photosynthetic activity within water by phytoplankton. Organisms have specific oxygen requirements. Low dissolved oxygen may prove fatal for many organisms for their survival. Oxygen absorbed from pot permagnate at 37 °C in 3 hrs. The value of absorbed oxygen of sample S_4 was 13.9 which is unpotable and remaining water samples are within the permissible limit. (Limiting range in drinking water is 0.1 according to Public Health Manual U.P.)

Chloride Content

Chloride is invariably present in small amounts in almost all natural waters and its content goes up appreciably with increasing salinity. Chloride usually occurs as NaCl, $CaCl_2$ and $MgCl_2$ in widely varying concentration in all natural waters. They enter water by solvent action of water on salts present in the soil from polluting material like sewage and trade wastes. The presence of chloride salts in drinking water in small amounts may not hurt us, but chloride by products can do it. High concentration of chloride can form different by product when mixed with organic matter. The value of chloride was recorded as 20 to 50 mg/L at all sampling locations. Chloride is corrosive and causes pitting in boilers.

Fluoride Content

Fluorides are essential for man and animal in small quantities, because it is a component part of bones and teeth. But excess of fluorine in water or air is poisonous because all fluorine compounds are poisonous. Fluorine poisoning destroys enamel of teeth and makes bones brittle. The value of fluoride content was recorded as 0.4 mg/L in all collected water sample.

Iron

Iron is one of the content of surface water causes staining problem and bitter taste above 2 mg/L. Iron bacteria (crenothrix) in the distribution system causing slime and objectionable order. Ferric and organically bound iron is the main source of iron. 0.5 mg/L iron was recorded in S_8 water samples and 0.8mg/L iron was recorded in S_2, S_5, S_7 water samples.

Nitrite

Nitrite content is found in surface water and pollution in the near past is source of it. Ingested nitrite reacts with secondary and tertiary amines to turns nitrosamine which may be carcinogenic. Only three water samples S_2, S_4, S_5 have nitrite content. The remaining water samples have not nitrite content.

Nitrate

The source of nitrate pollution is sewage pollution, use of fertilizers and leaching of nitrates from soil. It causes infantile methemoglobinemia and irritation of mucous membrane in adults. The result of nitrate content was as same as nitrite content.

Bacteriological Analysis of Water

Bacteriological analysis of water was done to differentiate the coliform group by presumptive, confirmed and completed coliforms tests.

Presumptive Coliform Test

Presumptive coliform test was done to confirm whether collected water samples contain lactose fermenting gas producing bacteria or not. The water samples collected from eight blocks of Banda region on Sept-2012 were subjected to presumptive test and most of the water samples showed positive results *i.e.* gas formation by changing the colour in the samples.

Confirmed Coliform Test

This test is used to confirm the presence of coliforms and to determine the MPN (Most probable number) value in water samples showing positive or doubtful presumptive test. All surface water samples gave positive confirmed coliform test and MPN value greater than 180. The MPN index[7] of ground water has been found to be lower than surface water.

Completed Coliform Test

By the experiments of completed coliform test it was observed that all eight surface water samples have large number of coliform colonies.

Centrifugation Test

By this experiment many types of bacteria have been confirmed. There were 3-4 types of bacteria have been examined under the microscope in each collected water sample.

Potable Water Quality Criteria

Quality criteria of water for drinking purposes have been prescribed by various authorities like World Health Organization (WHO), Indian Standards Institution (ISI) and Indian Council of Medical Research (ICMR) etc. Besides the chemical and physical criteria, these authorities have prescribed bacteriological standards for water as well. World Health Organization's guide lines for drinking water regarding bacteriological contents have been adopted by India[8].

Table 8.2: Bacteriological Analysis of Surface Water Collected from Eight Blocks of Banda Region on September 2012

Surface Water Sample	*Presumptive Coliform Test*			*Confirmed Coliform Test (MPN Index/ 100mL)*	*Completed Coliform Test (No. of coliform in 100ml)*		*Centrifugation Test*
	Single Tube	*Five Tube*					
	50 ml	*10 ml*	*1 ml*		*48 hrs at 37°C*	*72 hrs at Room Temp.*	
S_1 (Jaspura)	1	5	5	>180	15	20	Spiral, coccus, streptococcus, basilli
S_2 (Tindwari)	1	5	5	>180	10	15	Navicula, anavina
S_3 (Baberu)	1	5	5	>180	15	20	Basillus, spiral, coccus, spores
S_4 (Kamasin)	1	5	5	>180	15	20	Basillus, cocobasillus, navicula, diatom
S_5 (Mahua)	1	5	5	>180	10	15	Basillus, spores, coccus
S_6 (Visanda)	1	5	5	>180	10	15	Navicula, coccus, endospores
S_7 (Badokhar khurd)	1	5	5	>180	10	15	Navicula, coccus, spores
S_8 (Naraini)	1	5	5	>180	15	20	Navicula, diatom

WHO guidelines for drinking water points out the following criteria:

(a) Total coliforms: Zero per 100ml sample (0/100ml). Not more than 5 per cent of sample should be positive. The positive samples analysed for *E. coli* or feacal coliforms.

(b) *E. coli* : Zero per 100ml sample (0/100ml).

(c) Feacal coliform: Zero per 100ml sample (0/100ml).

Table 8.3: Indian Standard Specification for Drinking Water

Sl.No.	*Parameter*	*Requirement Desirable Limit*	*Remarks*
1.	Colour (Hazen units)	5	May be extended upto 50 if toxic substances are suspected.
2.	Turbidity (NTU)	10	May be relaxed up to 25 in the absence of alternate.
3.	pH	6.5 to 8.5	May be relaxed up to 9.2 in the absence
4.	Total hardness (mg/L)	300	May be extended up to 600
5.	Copper (mg/L)	0.05	May be relaxed up to 1.5
6.	Iron (mg/L)	0.3	May be extended up to 1
7.	Manganese (mg/L)	0.1	May be extended up to 0.5
8.	Chlorides (mg/L)	250	May be extended up to 1000
9.	Sulphates (mg/L)	150	May be extended up to 400.
10.	Nitrates (mg/L)	45	No relaxation
11.	Fluoride (mg/L)	0.6 to 1.2	If the limits below 0.6 water should be rejected, max limit is extended to 1.5
12.	Mercury (mg/L)	0.001	No relaxation
13.	Cadmium (mg/L)	0.01	No relaxation
14.	Arsenic (mg/L)	0.05	No relaxation
15.	Lead (mg/L)	0.1	No relaxation
16.	Zinc (mg/L)	5	May be extended up to 10.0
17.	Chromium (mg/L)	0.05	No relaxation
18.	Odour	Unobjectionable	–
19.	Taste	Agreeable	–
20.	Dissolved solids (mg/L)	500	2000
21.	Suspended solids (mg/L)	600	–

Experimental

Chemical Analysis

Temperature was measured by thermometer and for colour tests each sample was taken in a clean test tube and monitoring the colour visually in the presence of high intensity light. For determination of total suspended solids 100ml of the sample in a evaporating dish was placed in a oven for evaporation to dryness at 103-105°C. The dish was cooled down in desiccators and then weighed. With the help of a digital TDS meter the amount of dissolved solid was measured directly. By these two values the amount of total solid was determined easily. The pH values of different water samples were directly recorded by using a digital pH meter. The values of conductance of the different water samples were read out using a digital conductivity meter. The total hardness of water sample was determined by using total hardness eriochrome tablet which was taken in a porcelain dish and ammonia solution was added and then titrated by N/50 EDTA solution until red colour was changed into blue colour. Calcium hardness was calculated by taking 50ml water sample, 4N NaOH solution and calcium hardness eriochrome tablet in a conical flask which was titrated against N/50 EDTA solution until pink colour was changed into blue colour. Mg-hardness was calculated by subtracting value of total hardness and value of Ca-hardness. Turbidity in water sample was measured through digital turbidity meter. Alkalinity was determined by using 5 drops methyl orange indicator in 50ml water sample and it was titrated with N/50 sulphuric acid until the colour changes from yellow to rose red. For chloride content determination 50ml of water sample was taken in a porcelain dish and 5-6 drops of K_2CrO_4 was added and titrated with 0.02N silver nitrate solution with stirring till the brick red tinge appears. For determination of fluoride content 50ml water sample was taken in a test tube and then 5ml zirconylchloride octahydrate and 5ml alizarin zirconyl solution was added. The appeared colour was compared with fluoride standards. For testing of iron content 50ml of water sample, 2ml HCl and 1ml hydroxylamine hydrochloride was taken in a beaker which was heated upto 20-25ml. Then 2ml anthrophenophthalene and 5ml ammonium acetate was added into reaction mixture. The appeared colour was compared with Fe-standards. Nitrite content was determined by taking 50ml water sample in a Nessler's tube and then 1.5ml sulphinilic acid and 1.5ml naphthylamine was added. After 15 minutes pink colour was appeared to indicate the presence of nitrite content. For determination of nitrate content hot water bath method has been used. For determination of absorbed oxygen 100ml water sample and 10ml N/40 $KMnO_4$ solution was taken in a conial flask and keep it in incubator for 4hrs. After 4hrs 1g KI was added and then titrated with N/50 hypo solution. When solution become lighter in colour add starch indicator and again titrate it with hypo solution. The reading has been noted when blue colour solution of conical flask become decolourised.

Biological Test

Presumptive Coliforms Test

Lactose fermentation tubes were inoculated with different water volumes and production of acid and gas from the tubes was a presumptive evidence of coliforms in water samples.

Confirmed Coliform Test

This test was used to confirm the presence of coliform and to determine the MPN value in water sample showing positive or doubtful presumptive test. In the positive presumptive lactose broth tubes were inoculated into tubes of brilliant green lactose bile broth and inoculated at 37 °C for 48hrs and 72hrs at room temperature.

Completed Coliform Test

EMB agar plates were streaked from positive tubes with a sterile inoculating loop. Inoculated plates were incubated at 37 °C for 48 hrs and 72 hrs at room temperature in an inverted position. After 48 hrs and 72 hrs the inoculated plates were observed to detect coliform colonies.

Centrifugation Method

10ml water sample was taken in a centrifuge tube and centrifugate for 15 min in the centrifuge machine. Decant off supernatant and examined concentrated portion under the microscope to see types of bacteria.

Acknowledgement

The authors are thankful to Chitrakoot Dham Mandal Jal Sansthan, Banda and Faculty of Science and Environment, Department of Physical Science, Mahatma Gandhi Chitrakoot Gramodaya Vishwavidyalaya,Chitrakoot, Satna for providing necessary facilities in experimental analysis.

References

1. Wikipedia, the free Encyclopedia.
2. A. Gupta, Analytical Chemistry, 2010, 146.
3. S.A. Chaudhary and K.V. Sastry, J. Ecotoxicol. Environ. Monit, 2005, 15, 263.
4. D.L. Osmond, D.E. Line, J.A. Gale, R.W. Gannon, C.B. Knott, K.A. Bartenhagen, M.H. Turner, S.W. Coffery, J. Spooner, J. Wells, J.C. Walker, L.L. Hargrore, M.A. Foster, P.D. Robillard and D.W Lehing," water, soil and Hydro - Environmental Decision Support System," 1995, URL :www.water.ncsu.edu/watersheds/info/info/hmetals html.
5. V.Singh, S, K. Sharma and C.P. Singh, Handle Environment and Ecology, 2006, 245, 915.
6. R.P. Singh, J. L. Jethwa and B.B. Dhar, "Environmental Scenario in Mining Industry," eds. B.B. Dhar and N.C. Sexena Socioeconomic Impacts of Environment, 1997.
7. J.K. Hoskin, Public Health Report, 1934, 49, 393.
8. G. Panda and S. Gangopadhya, J. Indian Chem. Soc., 2012, 89, 939.

Chapter 9

Assessment of Ganges River Water Quality at Rani Ghat, Kanpur

Padma Saxena

Department of Zoology, D.A.V. College, Civil Lines, Kanpur – 208 001 Uttar Pradesh

ABSTRACT

Kanpur, Manchester of the East, famous for its leather industry, has its other side of story to tell too. It's also one of the most polluted cities in India as well as in World. In fact, environmental pollution has taken up by gigantic proportions and has become a major cause for concern among the population of the city. The main aim of the present study was to access the impact of urban and industrial activities or the water quality of Ganges river at Rani Ghat, Kanpur. For this Ganga river water samples were collected at Rani Ghat, Kanpur. The water samples were analyzed for temperature, pH, Acidity, dissolved oxygen (DO), biochemical oxygen demand (BOD), chemical oxygen demand (COD) and establish the relationship between concentrations of Chloride, Nitrate, Phosphate and Chromium in water at different periods during the year. Comparison between observed and estimated values based on physico-chemical parameters studies and water quality indices revealed that the river water quality at Rani ghat location were found to be contaminated.

Keywords: *Ganges river, Biochemical oxygen demand (BOD), Chemical oxygen demand (COD).*

Introduction

Over two third of the earth's surface is covered by water and less than a third is taken up by land. As the earth's population continues to grow, people are putting ever increasing pressure on the planet's water resources. Water is the basic need for every organism and life will cease to exist without water hence the study of water quality and water pollution is of utmost importance.

Today, over 29 cities, 70 towns, and thousands of villages along the Ganga river discharge over 1.3 billion liters sewage per day directly into the river, along with thousands of animal carcasses, mainly cattle. Another 260 million liters of industrial waste are added to this by hundreds of factories along the rivers banks. Municipal sewage constitutes 80 per cent by volume of the total waste dumped into the Ganga, and industries contribute about 15 per cent. The majority of the Ganga pollution is caused by organic waste, sewage, trash, food, and human and animal remains. Over the past century, city population by the Ganga has grown at a tremendous rate, while waste-control infrastructure has remained relatively non-existent[1]. Discharge of untreated wastewater from towns along Ganga which constitutes the major source of pollution load for the river, continues unabated. Water-related diseases are a human tragedy, killing millions of people each year, preventing millions more from leading healthy lives, and undermining development efforts[2]. About 2.3 billion people in the world suffer from diseases that are linked to water[3]. Some 60 per cent of all infant mortality is linked to infectious and parasitic diseases; most of them are water-related[4]. Providing clean supplies of water and ensuring proper sanitation facilities would save millions of lives by reducing the prevalence of water-related diseases. Thus, finding solutions to these problems should become a high priority for developing countries and assistance agencies. While water-related diseases vary substantially in their nature, transmission, effects, and management, adverse health effects related to water can be organized into three categories: water-borne diseases, including those caused by both fecal-oral organisms and those caused by toxic substances; water-based diseases; and water-related vector diseases.

In Kanpur, environmental pollution has taken up drastic proportions and has become a major cause for concern among the population of the city. Today, the river Ganga is a mere dumping ground for all industrial effluents and domestic sewage. Yet the people continue to drink and use this water as they have no other choice. In order to overcome such chronic problems the present work has been planned to know the present status of Ganga River at Rani ghat region in Kanpur.

Material and Methods

Rani Ghat is situated at 26.29 degrees North Latitude and 80.19 degrees East Longitude of river Ganga in Kanpur. The upstream of this sampling site receives organic load from different sources. Water samples were collected at different month for a period of years *i.e.* 2013-2014 between 7.00 AM to 5.00 PM from Rani Ghat in plastic jars, properly packed and transported to laboratory same day to avoid unpredictable changes in physico-chemical parameters. Standard method for the examination of water and waste water was used for analysis[5]. The sample for Cr estimation was collected in the glass bottles and preserved by adding 2 ml concentrated HCl. The temperature was recorded at the sites with the help of mercury thermometer and the DO water samples were fixed with the help of magnous sulphate and alkali-iodide-azide solution (2ml each) at the sites and analyzed in the laboratory using wrinkler´s modified iodide- azide method. The water samples were brought to the laboratory in ice boxes and subjected to the analysis of BOD (by incubating diluted sewage samples at 25°C for 5 days in dark) COD (by dichromate reflux method using a ferron indicator). pH using pH meter and acidity (by the Potentiometric titration method with the help of pH meter and diluted H_2SO_4 and NaOH, Chloride by Mohr's argentometric method using Potassium chromate as indicator), Phosphate (Stannous chloride method) and nitrate using the phenol – disulphonic acid colorimetric test[6]. For chromium concentration the sample was prepared by hot acid digestion followed by metal estimation through Atomic Absorption Spectrophotometer.

Results and Discussion

The results of samples collected at Rani Ghat, Kanpur in different seasons are given in Table 9.1.

Table 9.1: Psycho-chemical Properties of River Ganga at Rani Ghat

Parameters	*Season*		
	Summer ±SD	*Monsoon ±SD*	*Winter ±SD*
Temperature (°C)	24.73±2.46	18.25±9.85	15.17±1.53
pH	8.63±0.15	7.73±0.15	7.33±0.12
Acidity (mg $CaCO_3$/L)	64.56±3.59	64.53±1.10	58.77±0.85
Dissolved Oxygen (mg/l)	3.73±0.15	3.83±0.25	10.7±0.2
Biological Oxygen Demand (mg/l)	12.6±0.20	16.17±0.42	13.8±0.63
Chemical Oxygen Demand (mg/l)	71.23±9.70	76.03±1.72	55.63±0.83
Chloride (mg/l)	20.43±0.57	16.9±0.56	13.6±1.05
Nitrate (mg/l)	0.07±0.01	0.06 ±0.01	0.044 ±0.01
Phosphate (mg/l)	0.17±0.02	0.14 ±0.01	0.13 ±0.01
Chromium (mg/l)	0.24 ±0.01	0.19±0.01	0.18±0.01

The mean value of surface water temperature was observed to be 15.17±1.53°C in winter season, 24.73±2.46°C in summer season and 18.25±9.85°C in monsoon at Rani Ghat. Seasonal variation was recorded in the temperature. Highest values of temperature were recorded in summer followed by monsoon and winter [7–8]. The pH of a water body is very important in determination of water quality since it affects other chemical reactions such as solubility and metal toxicity[9]. During the course of study it was recorded that pH was always alkaline at all the seasons. The alkaline nature of river water may be due to high temperature which causes reduction insolubility of CO_2[10]. The seasonal average value of pH was 8.63±0.15 in summer, 7.73±0.15 in monsoon and 7.33±0.12 in winter (Table 9.1). The trend noticed in the present study was in accordance with the findings by other workers [11–12]. In general pH was within the limits of WHO standard value. Acidity of water is due to presence of carbon dioxide present in it. The seasonal average values of acidity in summer are 64.56±3.59, in monsoon 64.53±1.10 and in winter 58.77±0.85 (Table 9.1). Increase in acidity in summer and monsoon may be attributed to the high temperature and therefore high microbial activities or discharge of some acidic substances.

Dissolved Oxygen is an important parameter indicating the level of water quality and organic pollution in the water body[13]. The value of DO is remarkably significant in determining the water quality criteria of an aquatic system. In the system where the rates of respiration and organic decomposition are high, the DO values usually remain lower than those system, where the rate of photosynthesis is high[14]. In this study maximum DO was recorded, 10.7±0.2mg/l in the winter and minimum 3.73±0.15 mg/l in summer (Table 9.1). The present study reveals that there's a gradual decrease in DO from winter to monsoon. The value of DO during summer months is lower due to higher temperature and high rate of microbial decomposition of organic matter. Maximum values of DO in winter might be due to the fact that the solubility of DO increases with the decrease in water temperature[15].

The mean value of biological oxygen demand was observed 13.8±0.63 mg/l in winter season, 12.6±0.20 mg/l in summer season and 16.17±0.42 in monsoon. The biodegradation of organic materials exerts oxygen tension in the water and increases the biochemical oxygen demand during rains followed by winter and summer (Table 9.1). The mean value of chemical Oxygen demand was observed 55.63±0.83 mg/l in winter season, 71.23±9.70 mg/l in summer season and 76.03±1.72 mg/l in monsoon. The high and low value of BOD and COD during summer and monsoon season might be due to reduction in the volume of water and the addition of both organic and inorganic substances from tanneries, while the dilution of effluent by fresh water reduce the BOD and COD level[16]. Chloride concentration in water indicates presence of organic waste particularly of animal origin. The value of Chloride was observed 13.6±1.05mg/l in winter season, 20.43±0.57 mg/l in summer season and 16.9±0.56 mg/l in monsoon (Table 9.1). The values were highest in summer and lowest in monsoon and intermediate value were recorded in winter. The value of nitrate was observed 0.044 ±0.01 mg/l in winter season, 0.07±0.01mg/l in summer season and 0.06 ±0.01 mg/l in monsoon (Table 9.1). Higher nitrate values during summer might be due to increase in the degradation of organic matter by microbial activities. The value of phosphate were observed 0.13 ±0.01 mg/l in winter season, 0.17±0.02 mg/l in summer season and 0.14 ±0.01 mg/l in monsoon (Table 9.1). Similar findings have also been reported in Ghazipur for Ganga river[17]. Agriculture runoff containing phosphate fertilizers as well as the wastewater containing the detergents etc. tend to increase phosphate pollution in water. The value of chromium was observed 0.18±0.01mg/l in winter season, 0.24 ±0.01 in summer season and 0.19±0.01 in monsoon (Table 9.1).

As the facts suggests, we need to develop a scientific and sociological model with emphasis on scientific analysis and make people aware of the huge proportions of the environmental degradation of the water of the river Ganga, which flows through the city of Kanpur and is also the lifeline of the city.

References

1. Basant Rai (2013). Pollution and Conservation of Ganga River in Modern India *International Journal of Scientific and Research Publications* 3 (4): 1-4.
2. Tripathi G. (1983). Environmental problems of Indian Rivers and their ill effects. River Pollution and Human Health. R.S. Ambasht and B.D. Tripathi (Eds.), NECA, Varanasi: 23-33.

3. Singh Namrata (2010). Physicochemical properties of polluted water of river Ganga at Varanasi Int. J.Ener. Env.), 1(5) : 823-832
4. Khatoon N., Masihur R. and Khan A. H. (2013). Study of seasonal variation in the water quality among different ghats of river Ganga, Kanpur, India *J. Environ. Res. Develop.* 8 (1): 1-10.
5. American Public Health Association, APHA. (1998). Standard methods for examination of water and wastewater (20th Ed.). Washington, DC.
6. Jackson, M. L. (1958). Soil chemical analysis. Prentice-Hall, Englewood Cliffs, NJ.
7. Zafar A. and Sultana N. (2008). Seasonal analysis in the water quality of the river Ganga. *J. Curr. Sci.*, 12(1), 217-210.
8. Khanna D.R., Ashraf, J., Chauhan Beena and Bhutiani R. (2005). Physico-chemical analysis of Panv Dhoi River at Saharanpur (U.P.)., *Env. Conser. Jr.*, 6(2), 89-94.
9. Fakayode S.O. (2005). Impart of industrial effluent on water quality of the receiving Alero river in Ibadan, Nigeria, *Ajeam-Ragee*, 10(1),1-1.
10. Mahananda M. R., Mohanty B. P. and Behera N.R. (2010). Physico-chemical analysis of surface and groundwater of Bargarh district, Orissa, India, *IJRRAS*, 2(3), 284-295.
11. Dalal A. and Arora M. P. (2008). Seasonal physico-chemical fluctuations of water quality of river Hindon, *J. Exp. Zool. India*, 11(1), 125–131.
12. World Health Organization, (2008). Guidelines for drinking Water Quality : Vol. 1, *Recommend.* 2nd Edn.; *Geneva*, WHO, 1-10.
13. Wetzel R.G. and Likens G.E. (2006). Limnological analysis, 3rd Edn., *Springer-Verlag, NewYork*, 391.
14. Mishra A., Mukherjee A. and Tripathi B.D. (2009). Seasonal and temporal variation in physicochemical and bacteriological characteristics of River ganga in Varanasi, *Int. J. Environ. Res.*, 3(3), 395-402.
15. Zeb B.S., Malik A.H., Waseem A. and Mahmood Q. (2011). Water quality assessment of Siranriver, Pakistan, *Int. J. Phys.* Sci., 6(1), 7789-7798.
16. Alam Md JB, Islam MR, Muyen Z, Mamun M, Islam S (2007). Water quality parameters along rivers. *Int J Environ Sci Tech* 4: 159-167.
17. Yadav R.C, Srivastava V.C. (2011). Physico-chemical properties of the water of river Ganga at Ghazipur. *Indian J. Sci. Res.* 2(4) : 41-44.

Chapter 10

Plankton Diversity Studies of Upper Lake with Reference to Physico-chemical Status of Water

Sonam Gupta

Govt. M.L.B. Girls P.G College, Bhopal

ABSTRACT

The diversity of phytoplankton and its relationship to the physico-chemical characteristics was studied in Bhoj Wetland, Bhopal. Water sample from different sites of Upper Lake were collected and analyzed for their Limnological characters viz. Temperature, pH, Total hardness, Total dissolved solids, Dissolved Oxygen, BOD, COD, Total alkalinity, Chloride, Calcium, free Carbon dioxide, Sulphate, Nitrate, correlation between various water quality parameters. Phytoplankton is the base of several aquatic food meshes in a balanced ecosystem.

Keywords: Plankton, Upper lake, Physico-chemical status, Geo-position, Diversity.

Introduction

Water is a transparent fluid which forms the world's streams, lakes, oceans and rain and is the major constituent of living things. Water contains one oxygen and two hydrogen atoms with covalent bonds. Water covers about 71 per cent of the Earth's surface. Bhojtal is formerly known as Upper Lake is a large lake which lies on the western side of the capital city of Madhya Pradesh. It is a major source of drinking water for the city (Bajpai A, Dixit S and Verma N, 1993). Upper lake is situated on the west central part of Bhopal city and is surrounded by Van Vihar National Park on the south, human settlements on the east and north, and agriculture fields on the west. It has an area of 31 km^2 and drains a catchment or watershed of 361 km^2. It is the largest artificial lake in Asia. Water sample from different sites of Upper Lake were collected and analyzed for their Limnological characters *viz.* Temperature, pH, Total hardness, Total dissolved solids, Dissolved Oxygen, BOD, COD, Total alkalinity, Chloride, calcium, free Carbon dioxide, Sulphate, Nitrate, correlation between various water quality parameters. Phytoplankton is the base of several aquatic food meshes in a balanced ecosystem. There are two classes of plankton one is phytoplankton and other is zooplankton. Phytoplankton, also known as microalgae, are similar to terrestrial plants in that they contain chlorophyll and require sunlight in order to live and grow. Zooplanktons are small animals that occur in the water column of either marine or freshwater ecosystems. Abundance of zooplankton has great importance in fish farming (Bhuiyanet, 2008). Diversity of Plankton serves an important bioindicator to check the water quality of any aquatic habitat.

Study Area

The study area is situated on the catchment of the Upper Lake, Bhopal.

Sampling Station Near Sheetal Das ki Bagiya (S1)

Sheetal Das ki Bagiya is a great place to hangout, given the tranquility and beauty of the lake, especially during dawn and dusk. It is situated between 23"150.33 north and 77"74.744 east.

Sampling Station Near Lake View (S2)

Lake view site of Upper Lake is situated at 23"14.485 north and 77"22.559 east GPS location.

Sampling Station Near Garm Gaddha (S3)

This station of the Upper Lake is near karbala and situated between 23"15.560 north and 77"23.155 east.

Sampling Station Near Khanu Gaon (S4)

Khanu Gaon is situated between 23"15.630 north and 77"22.786 east and near VIP Road Bhopal.

Materials and Methods

Water Temperature

The water temperature was measured by using a probe and also using a mercury thermometer graduated up to 100°C with an accuracy of 0.1–0.2°C.

TDS

TDS was determined by TDS meter, which gives reading directly in mg/L.

pH

pH is the logarithm in base 10 of the reciprocal of the hydrogen ion concentration given in moles per liter. Hydrogen ion concentration (pH) was measured by Water Analysis Kit by using hydrogen ion selective electrode.

Dissolved Oxygen (DO)

Dissolved oxygen is the most significant parameter, because it regulates the metabolic processes of the organisms. The dissolved oxygen concentration depends on the physical, chemical and biological activities in the water body, and its measurement provides a good indicator of water quality. Changes in dissolved oxygen concentration can be an early indication of changing condition in the water body. Dissolved oxygen is measured by a number of methods including electronic oxygen meters but the most commonly used and preferred method for DO determination is the Winkler's method with azide modification. dissolved oxygen content was calculated by using the formula.

$$\text{Dissolved oxygen mg/L} = \frac{\text{mL of titrant} \times N \times E \times 100}{\text{Ml of sample}}$$

where,

N = Normality of titrant

E = Equivalent weight of oxygen

Total Alkalinity

The total alkalinity was obtained by adding carbonate and bicarbonate alkalinity. The carbonate alkalinity was determined by titrimetric method using phenolphthalein as Indicator the bicarbonate alkalinity was determined by titrating sample with standard acid solution using methyl orange indicator.

Total Hardness

Total hardness (as $CaCO_3$) was determined by EDTA tritrametric method using Erichrome Black-T indicator, values were computed by following formula:

$$\text{Total hardness mg/L} = \frac{\text{mL of titrant} \times 100}{\text{Ml of sample}}$$

Chloride

Chloride concentration was determined by argentometric titration method involving formation of reddish brown complex by adding Potassium Chromate which is titrated against Silver Nitrate solution.

Biochemical Oxygen Demand (BOD)

The biochemical oxygen demand (BOD) is an empirical test, in which standardized laboratory procedures are used to estimate the relative oxygen requirements of wastewater's, effluents and polluted waters. Biochemical oxygen demand was determined by measuring the difference of the oxygen concentration (By modified Winkler's method) between the sample and after incubating it for five days at 20°C.

Chemical Oxygen Demand (COD)

The Chemical Oxygen Demand (COD) is the amount of the oxygen consumed by organic matter from boiling acid potassium dichromate solution Reflux condensation method was used for the determination of Chemical Oxygen Demand.

Nitrate

Nitrate was determined spectrophometrically using the phenol disulphonic acid method on HACH DR 4000 UV-Vis spectrophotometer.Nitrite was analyzed using sulfanilamide method on HACH DR 4000 UV-VIS Spectrophotometer.

Phosphate

Phosphate was determined spectrophotometrically using the stannous chloride method on HACH DR 4000 UV-Vis spectrophotometer.

Sulphate

Sulphate was determined spectrophotometrically using the turbidimetric method on HACH DR 4000 UV-Vis spectrophotometer.

Calcium

Calcium was determined the EDTA titrimetric method

Plankton Identification

Process of collecting and interpreting representative portions of a resident aquatic community is most important to determine the community structure and function. For this, both phytoplankton and zooplankton of upper lake have been taken into consideration. Samples have been collected from the identification Adoni (1985) methods. Biological samples (phytoplankton) were preserved by adding 5 ml of 4 per cent formalin and the identification of phytoplankton was done with the help of standard works *viz.*, Edmondson (1965), Needham and Needham (1966). Plankton samples were collected by plankton net made of bottling silk cloth No. 25 (mesh size 0.064 mm) fortnightly.The standard method *i.e.* Drop count method for the analysis were followed, as prescribed in APHA (1995).

Results

Phytoplankton in Upper Lake

Distribution of phytoplankton species at different sampling stations in Upper lake during the study period.

Sheetal Das ki Bagiya (S1)

Total 25 species were recorded in which 18 species (sp.) belong to Chlorophyta, 4 sp. belong to Bacillariophyta 2 sp. Cyanophyta, Pyrrophyta 1 sp.

Lake View (S2)

Total 12 species were recorded in which 4 species (sp.) belong to Chlorophyta, 2 sp. belong to Bacillariophyta, 3 sp. Cyanophyta, Pyrrophyta 3 sp.

Garm Gaddha (S3)

Total 24 species were recorded in which 7 species (sp.) belong to Chlorophyta, 6 sp. belong to Bacillariophyta, 5 sp. Cyanophyta, Pyrrophyta 3 sp.

Khanu Gaon (S4)

Total 21species were recorded in which 7 species (sp.) belong to Chlorophyta, 6 sp. belong to Bacillariophyta 5 sp. Cyanophyta, Pyrrophyta 3 sp.

Zooplankton in Upper Lake

Sheetal Das ki Bagiya (S1)

Total 22 species were recorded in which 6 species (sp.) belong to Protozoa, 4 sp. belong to Rotifera and 5 sp. Cladocera, Copepoda 3 sp. Oligocheata, 2 sp. and Dipthera 2 sp.

Lake Biew (S2)

Total 21 species were recorded in which 6 species (sp.) belong to Protozoa, 5 sp. belong to Rotifera and Cladocera 2, Copepoda 4 sp., Ostracoda 2 sp. and Dipthera 2 sp.

Garm Gaddha (S3)

Total 16 species were recorded in which 5 species (sp.) belong to Protozoa, belong to Rotifera, 6 sp. Cladocera 3, Copepoda 2 sp.

Khanu Gaon (S4)

Total 20 species were recorded in which 3 species (sp.) belong to Protozoa, 6 sp. belong to Rotifera, 5 sp. Chirocephalidae 3 sp. Oligocheata 2 sp. and Dipthera 2 sp. Acartidae 1 sp.

Table 10.1: Phytoplankton Diversity of different Location of Upper Lake

Sampling Station	*Chlorophyta*	*Bacillariophyta*	*Cyanophyta*	*Pyrrophyta*	*Total*
S1	18	4	2	1	25
S2	4	2	3	3	12
S3	10	6	5	3	24
S4	7	6	3	5	21

Table 10.2: Physiological Status of Water of Upper Lake

Sampling Station	*Tem.*	*pH*	*Dissolve Oxygen (mg/ltr)*	*Alkalinity*	*Total Hardness*
S1	24.4	7.5	2.8 mg/l	2.5 mg/l	2.8 mg/l
S2	25.3	8.5	2.5 mg/l	3 mg/l	2.5 mg/l
S3	26.6	7	3.2 mg/l	2.1 mg/l	2.5 mg/l
S4	24.4	7.5	1.2 mg/l	2.6 mg/l	2.4 mg/l

Table 10.3: Chemical Status of Water of Upper Lake

Sampling Station	*Chloride*	*Free CO_2*	*Calcium*	*Nitrate*	*Phosphate*
S1	2.6	0.4	2.1	1.98mg/l	1.26mg/l
S2	2.5	0.2	4.3	1.80mg/l	1.23mg/l
S3	2.5	0.3	2.5	1.93mg/l	1.25mg/l
S4	2.4	0.4	2.2	1.90mg/l	1.34mg/l

Table 10.4: Chemical Status of Water of Upper Lake

Sampling Station	*Sulphate*	*Biological Oxygen Demand (mg/ltr)*	*Chemical Oxygen Demand (mg/ltr)*
S1	14 mg/l	3.8 mg/l	10.4 mg/l
S2	13 mg/l	3.5 mg/l	10.2 mg/l
S3	15 mg/l	2.6 mg/l	9.5 mg/l
S4	14.2 mg/l	3.1 mg/l	10.7 mg/l

Table 10.5: Zooplankton Diversity of different Location of Upper Lake

Zooplankton	*Sampling Stations*			
	S1	*S2*	*S3*	*S4*
Protozoa	6	6	5	3
Rotifera	4	5	6	6
Chirocephalidae	–	–	–	5
Oligocheata	2	2	–	2
Dipthera	2	2	–	2
Acartidae	–	–	–	1
Cladocera	5	2	3	–
Copepoda	3	4	2	1
Ostracoda	–	–	–	2
Total	22	21	16	20

Discussion

Water quality of this lake is changes during a decade are based on physico-chemical and biological analysis and they directly affect on the biodiversity of this lake (Agrawal 1993), (Das and Kalita 1990), (Hussainy 1965), (Iqbal *et al.*, 1984) Zooplanktons are indicator of productivity in water body. The abundance and growth of zooplankton varies with seasons and water properties. Composition and diversity of zooplankton provide information on the characteristics and quality of the water body (Adeyemi *et al.*, 2009; Okayi *et al.*, 2001). Kedar *et al.* (2008) and Dhanapati (2000) reported that rotifers are chiefly freshwater forms and abundance of these organisms in water body can be related to favourable temperature and availability of

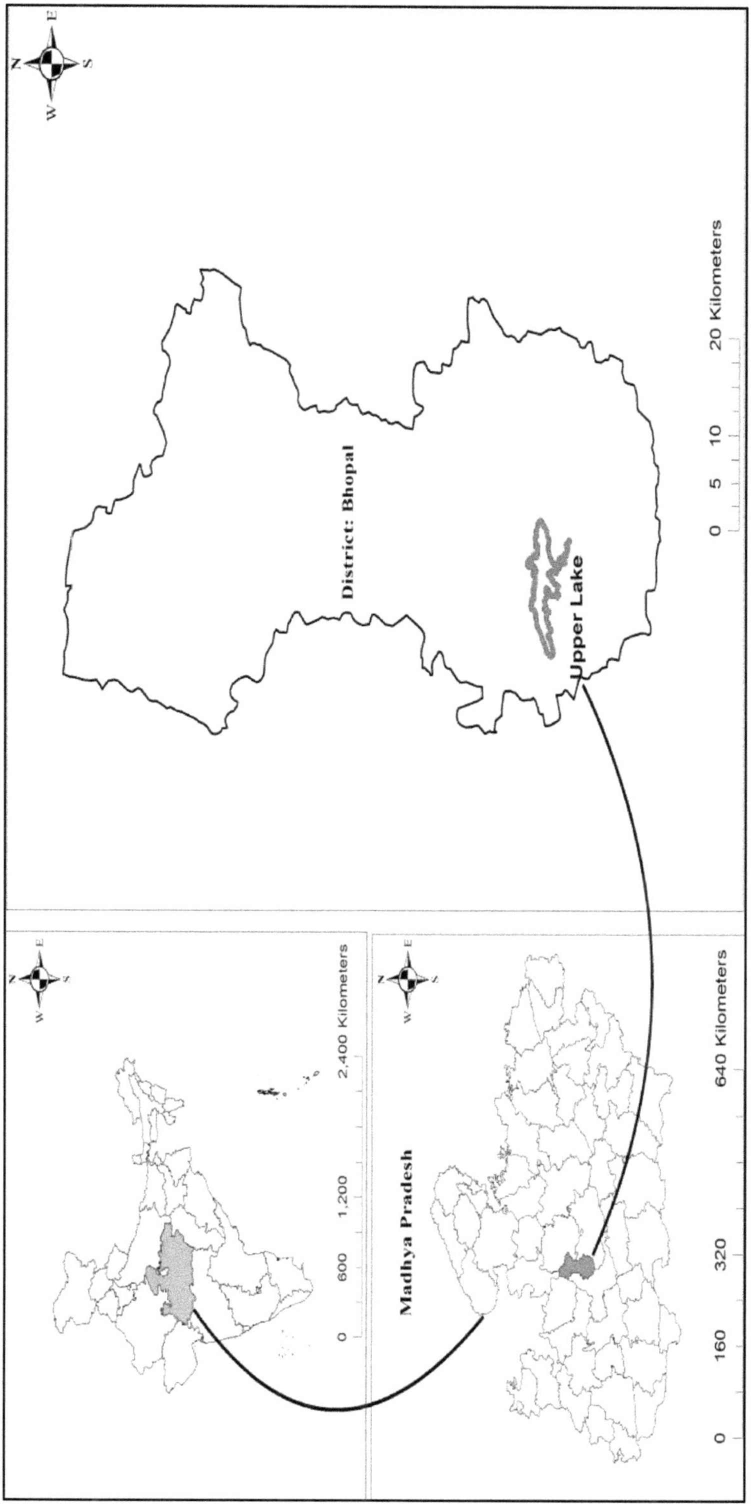

Figure 10.1: Position of Upper Lake on Map of India.

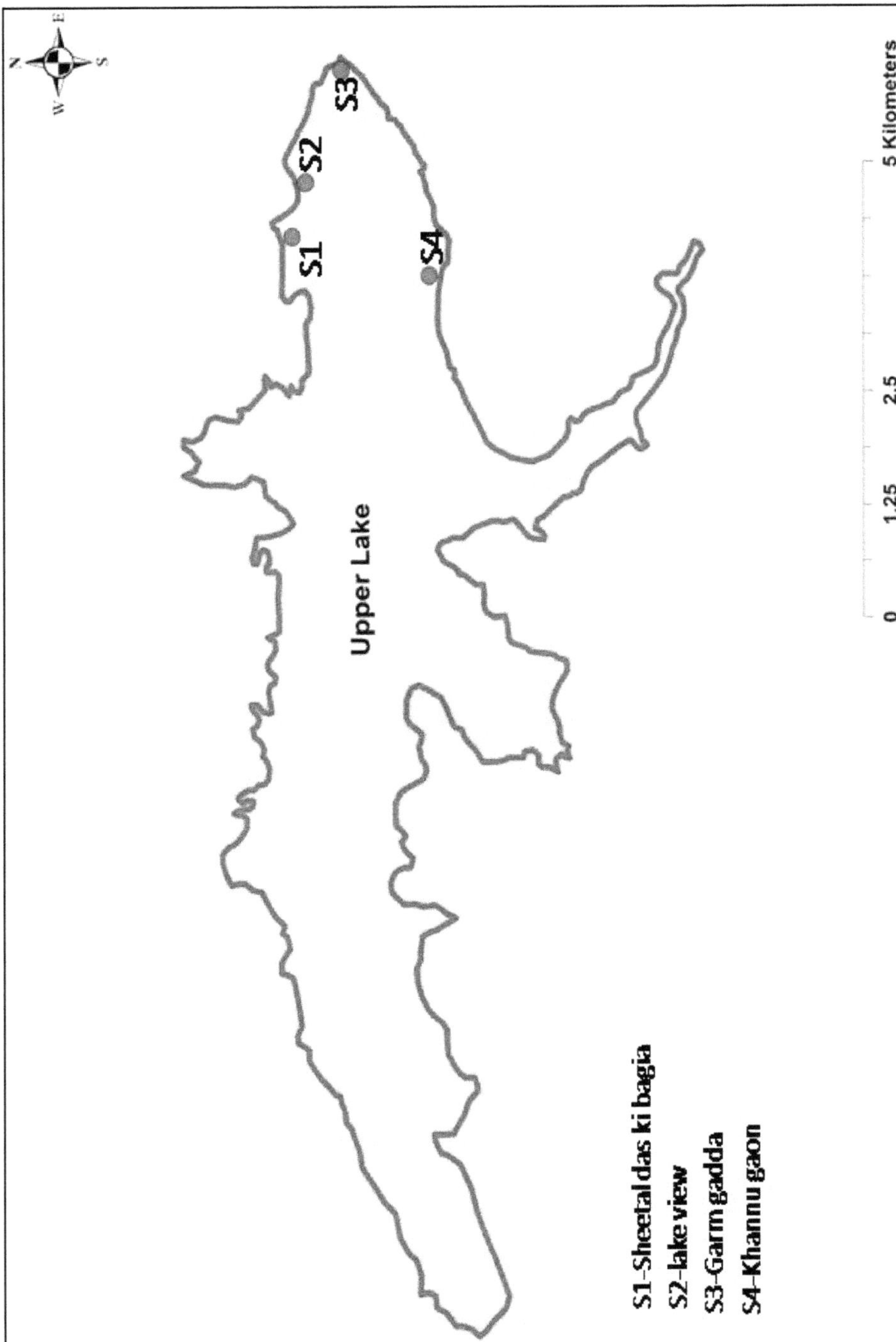

Figure 10.2: Outer Boundary Drawing of Upper Lake Bhopal with Sampling Station in Dotted Point.

abundant food in the form of bacteria, nanoplankton and suspended detritus which is suitable conditions for their survival. The presence of *Brachionus* spp., *Keratella* spp., *Bosmina* spp., *Mesocyclops* spp. and *Daphnia* spp. in all the three lakes indicates the higher trophic status of these lakes as these species are indicator of eutrophication (Kumar *et al.*, 2011; Sharma, 1983; Chaurasia and Adoni, 1985; Agarkar *et al.*, 1994; Wanganeo and Wanganeo, 2006; Kumar *et al.*, 2010).

Acknowledgement

The authors are thankful to Dr. Rashmi Vyas (Guide) and Dr. Suchetra Benarge (Head of the Department) for their valuable suggestion and facilities provided during this work.

References

1. A. Kumar, B.S. Bisht, V.D. Joshi, A.K.Singh, A.Talwar, Physical, Chemical and Bacteriological study of water from Rivers of Uttarakhand. *J Hum Ecol.*, 32(3), 010).
2. A. Wanganeo, R.Wanganeo, Variation in Zooplankton population in two morphologically dissimilar rural lakes of Kashmir Himalayas. *Proc. Nat. Acad. Sci. India*, 76 (B) III (2006).
3. Adoni AD (1985). Work Book of Limnology, Pratiba Publication Sagar, (M.P), India.
4. Agrawal, (1993) A. studies on physico-chemical and Biological Characteristics of river Betwa from Nayapura to Vidisha, Ph. D. Thesis (Chemistry) BU, Bhopal (1993).
5. APHA (1995). Standard methods for the examination of water and waste water, 19th edition American Public Health Association Washington DC.
6. B.K.Sharma, The Indian species of the genus Brachionus (Eurotatoria monogononta: Brachionidae). *Hydrobiologia*.104, 31-39 (1983).
7. Bajpai A, Dixit S and Verma N (1993). Limnological studies to assess water quality of upper lake, Bhopal. *National Seminar on Conservation and Development of Acquired Resource* 23-24.
8. Bhuiyan, M. T. Islam, R.Sharmeen, (2008). Occurrence and abundance of some copepods in a fish pond in Rajshahi, bangladesh in relation to the physico-chemical conditions. *J. bio-sci.*, 16, 115-119.
9. Das, H. B. and Kalita, H. (1990). Physico- chemical quality of water, Mizoram JIWWA, 22(2): 203-3.
10. Edmondson WT (1959). Freshwater Biology. 2nd ed. John Wiley and Sons, Inc. New York, London.
11. G.T. Kedar, G.P. Patil, S.M. Yeole (2008). Effect of Physico-Chemical Factors on the Seasonal Abundance of Zooplankton Population in Rishi Lake. *Proceedings of Taal 2007*: the 12th World lake Conference: 88-91.

12. Hussainy, S.U. (1965). Limnological studies of the departmental pond at Annamalainagar. Environ Health. 7: 24-31. *International Journal of Lakes and Rivers*,Vol. 2, No. 1, 37-44 (2009).

13. Iqbal, S. A. (1984) Khan, S.S.; Chaghtai, S.A. ;Hussain, I. Assessment of pollution levels of river Betwa. *J. Sci. Res.* 6(3): 165-170.

14. M.S. Agarkar, H. K. Goswami. S.Kaushik, S. N. Mishra,A.K. Bajpai, U.S.Sharma, Biology, conservation and management of Bhoj wetland 1. Upper lake Ecosystem in Bhopal. *Bionature*, 14 (2), 1-119 (1994).

15. M.V.S.S.S.Dhanapati, Taxonomic notes on the Rotifers from India – IAAB Publication, Hyderabad– 175 (2000).

16. Madhu Verma M. (2001). Economic Valuation of Bhoj Wetland for Sustainable Use. Project undertaken for the World Bank Aided India : Environmental Management Capacity Building Technical Assiatance Project Implemented by the Ministry of Environment and Forest and Coordinated by EERC Implementation Cell at IGIDR, Mumbai.

17. Needham GT and Needham PR (1966). A Guide to Freshwater Biology, 5th Edition, Holden Day Inc Sanfransisco 108.

18. R.G. Okayi, C.Y. Jeje, F.O. Fagade, Seasonal patterns in the zooplankton community of River Benue (Makurdi), Nigeria. *African Journal on Environmental studies*, Vol. 2, No. 1, 9-19 (2001).

19. S. K.Chaurasia, A. D. Adoni, Zooplankton dynamics in a shallow eutrophic lake. *Proc. Nat. Symp. Pure and Appl. Limnology (ed) Adoni*. A. D. Bull. Bot. Soc Sagar, 32, 30-39 (1985).

20. S.O. Adeyemi, I.A. Adikwu, P.M. Akombu, J.T. Iyela. Survey of Zooplanktons and Macroinvertebrates of Gbedikere Lake, Bassa, Kogi State, Nigeria.

21. V. B. Kumar, V. K. Pradeep, S.V. Roopa, Aquachemistry, Zooplankton and Bacterial Diversity in three Ponds of Karwar District, Karnataka. *Recent Research in Science and Technology*, 3 (4), 39-48.

Chapter 11

Quality Assessment of Groundwater in some Rural Areas of Kanpur

Atul Kumar Misra

Department of Zoology, D.A.V. P.G. College, Kanpur, U.P.

ABSTRACT

Water Pollution is one of the most serious problems of living organisms. Kanpur is well known thickly populated industrial city of Uttar Pradesh. Water is the biggest and chief natural resource, necessary for conservation for survival of living organisms on earth. It is the biggest necessity for life and in developing countries like India ground water is the most important source for drinking, irrigation, and industrial purposes. But unfortunately due to wide spread over use of harmful chemicals in agriculture, industrial effluents, pollution of rivers etc. ground water is getting increasingly contaminated with pollutants.

A laboratory study was conducted for the quality assessment of ground water in some selected rural areas of Kanpur. For the experimental purposes, four water samples were collected from different locations of the city in the month of September 2014 (Two open well and two bore well). The results were analyzed with standard value prescribed. It was concluded that the sampling sites No. A to C showed that all the physico-chemical Parameters within the range of standard water quality as prescribed by WHO and ISI 10500-91 physico-chemically, the quality for drinking purpose. The results of present study also concluded that most of the physico- chemical parameters are within permissible limits except samples No. D.

Keywords: *World Health Organization (WHO), Physico-chemical analysis, Industrial effluents, Physico-chemical parameters.*

Introduction

Groundwater is ultimate, suitable and safest water resource for living organisms on earth because it is a balanced concentration of the salts and minerals, which is required for human consumption. In another words water is the biggest necessity for survival of plants and animals. Water is very essential factor for the living organisms. In water many salts, minerals and nutrients are suspended in balanced forms. Drinking water is obtained from a variety of sources, depending on availability of surface water like lakes, rivers, reservoirs, ponds, pools and ground water (Nagpal *et al.*, 2004; Baig *et al.*, 2009).

In India, most of the population is totally dependent for their survival on ground water because it is the main source for drinking and survival of life on earth. In many developing countries like India, ground water is the most important source for drinking, domestic and irrigation purposes. People in the city as well as rural areas are totally dependent on ground water for their domestic and occupational requirements. It is well known fact, that ground water is comparatively much better, clean, safe and fit free from pollution in comparison of surface water, but in city and rural areas discharge of industrial effluents, solid waste dump and sewage caused the ground water get polluted by the many anthropogenic activities of human beings and created so many health problems among living organisms. In present situation, over burden of population, unplanned way of urbanization and dumping of polluted water at inappropriate place caused the discharge of harmful and toxic compounds to the ground water and it caused serious problems in human beings (Todd, 1980; Raja and Sharmila, 2002; Mahanta *et al.*, 2004; Baig *et al.*, 2009).

The physico-chemical characters of water can be changed due to presence of different pollutants and nutrients through different sources like, industrial effluents from tanneries and leather factories around the city, sewage and agricultural by products (Zatis, 1971; Olimax and Sikorska, 1975). It is observed, that these effluents and sewage water are not treated and purified before their domestic uses, such type of ground water is very dangerous and harmful for human consumption as well as for domestic and cattle uses. Now-a-days, over burden of population, dumping of polluted water in an open areas, and unplanned way of urbanization caused the discharged of harmful toxic chemicals in the ground water and many impurities suspended in water and it caused serious problems in living organisms and human beings also (Chilton, 1996; Mahanta *et al.*, 2004).

It is well known fact, that among the various sources of water, ground water is very safest and easily approachable source for drinking and domestic purpose but quality of ground water is influenced by many factors, which are responsible for the impurity of ground water in the natural conditions (Todd, 1980; Forstner and Wittman, 1981; Tiwari and Dhar, 1994; Canter and Sobatini, 1995 and Singh *et al.*, 2009). In present situation, the importance of ground water for survival of human, population can not be ignored. In India, specially in Uttar Pradesh, there are many thickly populated cities where more than 80 per cent population are dependant on ground water for drinking and other domestic purpose (Ramachandraian, 2004, Singh *et al.*, 2009).

The objective of the present study is to assess the quality of ground water for drinking purpose of some selected rural areas of Kanpur Region of Uttar Pradesh. The investigation is also a possible attempt to carryout to study, the physico-chemical analysis of drinking water from the selected areas on the basis of their physico-chemical parameters of experimental samples in the laboratory analysis.

Materials and Methods

Study Area

The experiment was conducted in the month of September, 2014 at Department of Zoology, D.A.V. College, Kanpur. In the Kanpur city, the hottest months are April to mid June. In these months, the temperature varies from 28°C to 43°C in summers, while the coldest months are December and January. The average annual rainfall in the city, between 900 mm to 1350 mm in the rainy season. The mansoon months are July to mid September. During the post mansoon period, the rate of humidity is very high. In the summer season, level of ground water decreased due to requirement of water by human beings. The quality of ground water in the study area is badly affected due to industrialization in the city. Due to over burden of population, improper drainage system, and many anthropogenic activities by human beings, the quality of ground water is badly affected and impured in the study area.

Preparation of Water Samples

Ground water samples were collected in the month of September, 2014. These samples were collected from two open well (Depth 35-65 feet). Similarly other water samples were collected from two bore well (Depth 185-250 feet). Open well samples and bore well samples were coded OW-S and BW-S respectively. The water samples were collected in sterilized screw-capped plastic canes of two liters as per standard procedure of collection prescribed. Just after the collection, the samples were stored at low temperature (5°C) to prevent vaporization and biodegradation before their physico-chemical analysis. Water samples from four sampling sites (two open well and two bore well) were collected at different places in the rural areas of the Kanpur Region.

The sampling location and their sources are shown in the Table 11.1.

Table 11.1

Sample No.	*Sampling Location*	*Source*
OW-S-(A)	Kalyanpur-Bithoor Road	Open Well
OW-S-(B)	Rawatpur Gaon	Open Well
BW-S-(C)	Kalyanpur-Bithoor Road	Bore Well
BW-S-(D)	Rawatpur Gaon	Bore Well

Quality Assessment of Groundeater in some Selected Rural Areas of Kanpur

The collected water samples were physico-chemically analyzed in an ideal laboratory conditions. Total fourteen physico-chemical parameters are analyzed.

These parameters are Colour, Odour, pH-value, Electrical Conductivity, Total Dissolved Solids, Total Hardness, Alkalinity, Calcium, Chloride, Nitrate, Turbidity, Sodium, Potassium and Magnesium. The physico-chemical parameters were estimated as standard methods (A.P.H.A.,1998). To ensure accurate, analysis was done in triplicates and for the results, mean value was taken for consideration. The results were compared with the prescribed guideline given by W.H.O. (1997) and Indian Standard value (I.S.105005-91).

Results and Discussion

The physico-chemical analysis, to test and assess, the water quality of ground water in various places has been carried out for the results. The status of water quality of ground water in same rural areas of Kanpur at different places is presented in Table 11.2.

Table 11.2: Showing the Physico-chemical Analysis of Water Sample A, B, C, D, E and F from different Localities of Kanpur City

Sl.No.	*Physico-Chemical Parameters*	*Sampling Sites*				*WHO 1997*	*ISI 10500-91*
		A	*B*	*C*	*D*		
1.	Colour	Colour Less	Colour Less	Colour Less	Colour Less	–	–
2.	Odour	Odour Less	Odour Less	Odour Less	Odour Less	–	–
3.	pH Value	7.3	7.5	8.0	9.0	7.0-8.5	6.5-8.5
4.	Electrical Conductivity (E.C.)	1325	450	1175	2400	1400	–
5.	Total Dissolved Solids (T.D.S.)	470	370	520	1225	1000	500
6.	Total Hardness (T.H.)	290	440	360	885	500	300
7.	Alkalinity	142	125	186	232	120	200
8.	Calcium (Ca^{++})	35.2	76.5	65.8	130.6	100	75
9.	Magnesium (Mg^{++})	45.80	148.60	75.60	148.6	150	30
10.	Chloride (Cl^{-})	130	245	185	280	250	250
11.	Nitrate (NO_3^{-})	18.4	17.4	40.2	28.7	45	100
12.	Turbidity	0.3	4.5	0.1	0.5	5.0	10
13.	Sodium (Na^{+})	375.5	268.4	160.8	194.6	200	200
14.	Potassium (K^{+})	4.838	0.425	3.800	1.008	–	–

1. Most parameters are represented in mg/L except pH value; 2. Turbidity = N.T.U.; 3. Electrical Conductivity (E.C.) = Micromho/cm; 4. World Health Organization (W.H.O.); 5. Indian Standard IS-105500-91.

pH Value

The term pH is used to express, the intensity of alkalinity or acid condition in a solution. The pH values of water samples varied upto 7.3 to 9.0. The pH value in the water samples within the prescribed limit. Although pH is not harmful as such but if its value is more than 11, it can cause irritation and skin disorder in human beings (Khan *et al.*, 2013). Most of the water samples are slightly alkaline due to the presence of carbonates and bi-carbonates.

Electrical Conductivity (E.C.)

Electrical conductivity is a measurement of water capacity to carry electric current. The (E.C.) values were recorded in between 1400 as prescribed by W.H.O. The E.C. range were 450 micromoles/cm to 2400 micromoles/cm in the sampling sites (B) and (F), because very high amount of dissolved inorganic substances are dissolved in ionized form.

Total Dissolved Solids (T.D.S.)

The salinity of ground water is denoted by the Total Dissolved Solids (T.D.S.). The sampling site (D), higher range of T.D.S. recoded. It is obvious, that water containing more than 500 mg/L of T.D.S. is not suitable for drinking purpose.

Total Hardness (T.H.)

According to Trivedi and Goel (1986), total hardness is the property of water which shows the leather formation with soap and it also increased, the boiling point of water. The hardness values were recorded, which ranges from 290 mg/L to 885 mg/L. It is also recorded that in sample No. (D), the hardness were higher than the prescribed limit.

Alkalinity

Alkalinity is most important chemical parameter of quality assessment of water. It is a capacity to neutralize a strong acid into its normal form due to the presence of carbonate, bi-carbonate and hydroxide compound of Sodium, Potassium, and Calcium. Generally, the alkalinity of water itself is not harmful, but large amount of alkalinity have bitter taste of water. The values of total alkalinity of water samples were recorded 125 mg/L to 232 mg/L.

Calcium (Ca^{++}) and Magnesium (Mg^{++})

It is well known fact, that the presence of calcium and magnesium in sample is directly related to the hardness of water. The presence of Calcium ranged between 35.2 mg/L to 130.6 mg/L found below the prescribed normal range.

The presence of Magnesium content in water samples were collected from the various sites ranging from 45.80 mg/L to 148.60 mg/L within the normal limit.

Chlorides (Cl^-)

In the present study, chloride concentration were recorded 130 mg/L to 280 mg/L. Higher range of chloride concentration were recorded in the industrial area sites, *i.e.* sampling site (D) due to discharged sewage near the sampling sites.

Nitrate (NO_3)

The presence of Nitrate is very important for suitability for drinking purpose. In the sampling sites, the nitrate contents of the samples were present within the permissible limits.

Turbidity

The turbidity in the ground water varied from 0.1 to 4.5 NTU. This range is below and normal value found within the limits prescribed by (W.H.O. and IS-10500-91).

Sodium (Na^+)

The presence of sodium were recorded between 160.8 mg/L to 375.5 mg/L. But sample (A), (B) and (C) showed higher value than the prescribed limit.

Potassium (K^+)

In fresh water, the chief source of Potassium by the mineralization of rocks, but due to disposal of waste water, the quantities of Potassium (K^+) increase. The presence of Potassium in the water samples varied from place to place as presented in Table-II.

In general, the water quality of ground water in rural areas of Kanpur City is not totally fit and suitable for drinking purpose because, many chemical parameters like Electrical Conductivity, pH Value, Alkalinity of Water, Total Hardness and presence of Calcium, Magnesium, Sodium, Potassium, Chloride and Nitrates ranges over then prescribed limits. Most of parameters in various samples fluctuate due to seasonal variations in the nature. Present study indicates that, the extent of pollution in water due to industrial discharge, urbanization, and other anthropogenic activities by human interventions in the ground water quality. It is also concluded that, water quality of ground water in Kanpur is not totally fit and suitable and causes so many health problems among human beings.

Acknowledgements

I am very thankful to, Principal, D.A.V. College, Kanpur and Head, Department of Zoology, D.A.V. College, Kanpur for providing all the required facilities, kind support and cooperation to carry out study.

References

1. American Public Health Association (A.P.H.A.) (1998) : Standard Methods for the examination of water and waste water. Washington, D.C.
2. Baig, J.H.; Kazi, T.G.; Arain, M.B.; Afridi, H.I.; Khandhro, G.A.; Sarfraj, R.A.; Jamal, M.K. and Shan, A.Q. (2009) : Evaluation of Arsenic and other physico-chemical parameters of surface and ground water of Jamshoro, Pakistan. *Journal of Hazardous Materials* : 166, 662-669.
3. Canter, L.W. and Sabatini, D.A. (1995) : Contamination of Public Ground Water Supplies by Supefund Sited. *International Journal of Environment Studies*, 46 : 35-67.

4. Chilton, J. (1996) : Water Quality Assessments - A Guide to use of Biota, Sediments and water in Environmental Monitoring, 2nd Ediction (Ed. D. Chapman), E and F.N. Spon, London.

5. Forstner, U. and Wittman, S. (1991) : Integrated Pollution Control, Springer, Verlog. Berlin, Germany.

6. Khan, I.; Javed, A. and Khurshid, S. (2013) : Physico-chemical analysis of surface and ground water around Singrauli Coal Field, District Singrauli, Madhya Pradesh, India. *Environ. Earth Sci.* 68, 1849-1861.

7. Mahanta, B.N.; Sarkar, B.C.; Singh, G.; Saikia, K.; Paul, P.R. (2004) : Multivariate statistical modeling and indexing of ground water quality in and around Jharia Coal Fields. Jharkhand. N.S.E.E.M.E.

8. Nagpal, A.K.; Singh, S. and Sambyal, V. (2014) : Physico-chemical analysis of ground water of Village-Jalalabad, Distt, Taran-Taran, Punjab : *Journal of Biospectra*, Vol. : 9(1), 125-128.

9. Olimax, T.; Sikorska, U. (1975) : Field experiment on the effect of municipal sewage on macrophytes and epifauna in the lake littoral. *Bull. Acad Pol. Sci. Clii* : 445-447.

10. Raja, R.E.; Sharmila, L.; Princy Merlin, J. and Chritopher, G. (2002) : *Indian J. Environ. Prot.* 22(2), 137.

11. Ramachandriah, C. (2004) : Right to drinking water in India Centre for Economic and Social Studies : 56-46-58.

12. Singh, A.K.; Verma, K. and Sharma, D.K. (2009) : Quality assessment of drinking water in the vicinity of domestic waste water drains of Amas Block Bazar. *Proc. Zool. Soc.* India 8(2) : 113-121.

13. Tiwary, R.K. and Dhar, B.B. (1991) : Environmental Pollution from Coal Mining Activities in Damdar River Basin, India Mine Water and the Environment, 13, 1-10.

14. Todd, D.K. (1980) : "Ground Water Hydrology", London J. Wiley and Sons.

15. Trivedi, R.K. and Goel, P.K. (1986) : Chemical and Biological methods for water pollution studies, Environmental Publication, Karad.

Chapter 12

Vulture Identification, Conservation and Awareness in Tikamgarh District, Madhya Pradesh

Jagdish Prasad Rawat

Department of Forest, Government of Madhya Pradesh, Tikamgarh, Madhya Pradesh

ABSTRACT

Vulture has an important role in the natural environment. If it weren't for vultures, many parts of the world would look like junkyards of bones and rotten meat. They are nature's disposal squads or "incinerators". Vultures provide the society with a number of services, most notably disposal of carrion. Vultures also help to control livestock disease as brucellosis, tuberculosis and anthrax by disposing of infected carcasses. The dying out of the vultures would be an irreplaceable loss of a link in the food chain. They come to the rescue of man to dispose of carcasses when animals die during natural calamities such as floods, storms, drought and war. By cleaning carcasses of dead animals, they act as the sanitation department of the natural world. These services have an impact on human health, economic activity and on environmental quality. Vultures are important for their considerable cultural and religious significance that some communities attach to their role of disposing of human bodies, such as followers of the minority Parsi faith.

Introduction

A Vulture is a scavenging bird of prey. There are 22 kinds of vultures which include 15 types of old world vultures and 7 types' new world vultures. The old world vultures belong to the family Accipitridae, while new world vultures belong to family Cathertidae. The scavengers have bare head and necks *i.e.* featherless. The beak are longed and hooked to feed upon the carcasses. The toes and tarsi are also featherless, thus helping in quick clean up after feeding. The unique feature is that they take a bath after each meal so that the blood from carcasses does not stay in their feathers or on their skin.

Vulture has an important role in the natural environment. If it weren't for vultures, many parts of the world would look like junkyards of bones and rotten meat. They are nature's disposal squads or "incinerators". Vultures provide the society with a number of services, most notably disposal of carrion. Vultures also help to control livestock disease as brucellosis, tuberculosis and anthrax by disposing of infected carcasses. The dying out of the vultures would be an irreplaceable loss of a link in the food chain. They come to the rescue of man to dispose of carcasses when animals die during natural calamities such as floods, storms, drought and war. By cleaning carcasses of dead animals, they act as the sanitation department of the natural world. These services have an impact on human health, economic activity and on environmental quality. Vultures are important for their considerable cultural and religious significance that some communities attach to their role of disposing of human bodies, such as followers of the minority Parsi faith.

Tikamgarh District is located in the northern part of Madhya Pradesh. It forms the north-western part of Sagar District. It lies on the Bundelkhand Plateau between the Jamni, a tributary of Betwa and Dhasan rivers. It extends between the latitude 24 degree 26 minute and 25 degree 34 minute N and between 78 degree26 minute and 79 degree 21 minute Longitudes. The shape of district is triangular. The northern margin is very irregular. The maximum length of the district is about 119 Km. From North to South and width about 80 Km. Tikamgarh District is bounded by Chhatarpur district

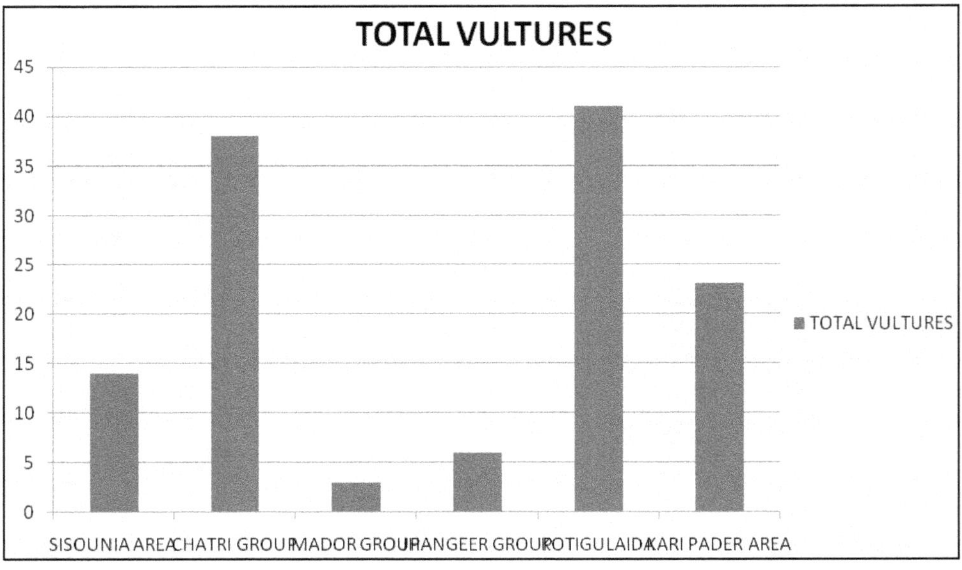

Figure 12.1: Area-wise Total Vulture Population Reported during the Census in Tikamgarh.

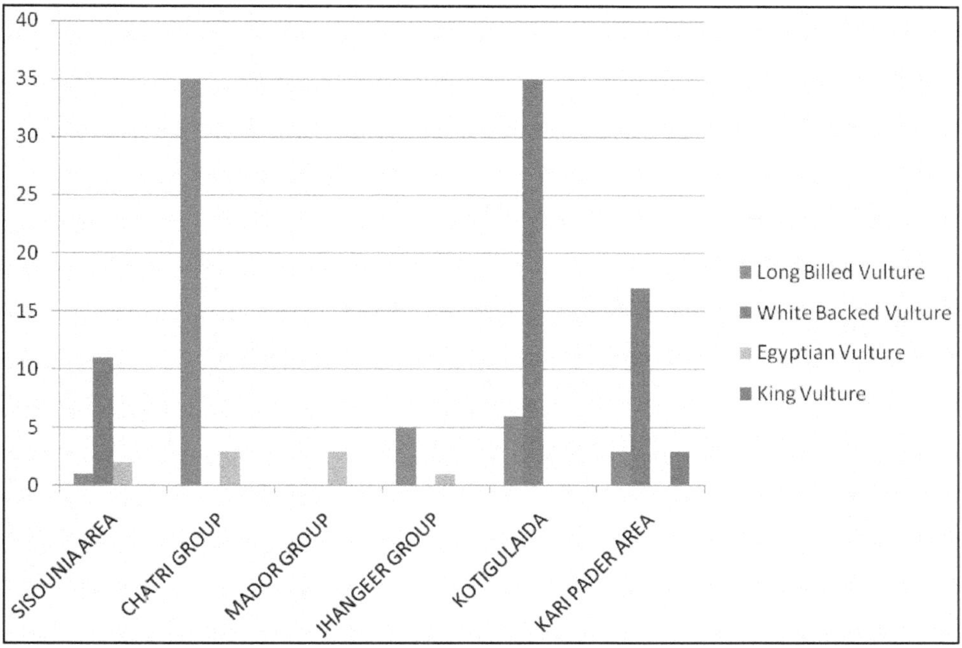

Figure 12.2: Area-wise Total Vulture Species Reported during the Census in Tikamgarh.

to east, Lalitpur district Uttar Pradesh to West, Jhansi to North and Sagar to South. The western and Eastern boundaries are formed by two big rivers.

Tikamgarh abodes a promising vulture population with suitable habitat. For the conservation of vultures that are critically endangered, it is important to protect all the breeding sites as well as feeding and roosting sites. From this point of view it is significant to create awareness through workshops among the people of Tikamgarh from all fields so as to protect the remaining vulture population. It is also vital to know the exact population of vultures in the region so as to proceed with the conservational measures. Therefore a workshop as well as census was conducted in Tikamgarh with the combined efforts of Forest Department, Tikamgarh and Department of Zoology, University of Lucknow. Experts from Gujarat, Satna, Jhansi, Ujjan were included.

The census result showed a total of 125 vultures and 39 nests. The maximum vultures were that of White Backed Vultures (63) followed by Long Billed Vultures (50) and Egyptian Vultures (9). The least number were that of King Vultures (3). No juveniles were reported during the census. The maximum vultures were reported in Kotigulainda (41) followed by Chatri area (38). The minimum vulture population was reported in Jahangeer area and Maror area *i.e.* 6 and 3 respectively. The census result and photographs of vulture species reported are attached as Annexure-1 and 2

Table 12.1: Enuneration of Vultures in Tikamgarh Districts–Year 2012

Sl.No.	*Area*	*Vulture Species*	*Number*	*Gyps Indices (Long Billed Vulture)*	*Gyps Bengalese's (White Backed Vulture)*	*Neoprene percnopterus Egyptian Vulture*	*King Vulture (Sarcogyps calvus)*	*No. of Nests*	*Total Vulture*
1.	Sinoniya	Long Billed vulture	01	01	11	02	0	01	14
		Egyptian Vulture	02						
		White backed vulture	11						
2.	Cenotaps group	Long Billed vulture	35	35	0	03	0	14	38
		Egyptian Vulture	03						
3.	Mador	Egyptian Vulture	03	0	0	03	0	0	03
4.	Jahngeer Fort group	Long Billed vulture	05	05	0	01	0	01	06
		Egyptian Vulture	01						
	Koti gulenda	Long Billed	06	06	35	0	0	0	41
		White backed vulture	35		35			15	
	Kari padar	Long Billed	03	03	17	0	03	0	23
		King Vulture	03				03		
		White backed vulture	17						
		G.Total		50	63	09	03	39	125

Long Billed vulture (*Gyps indicus*), White backed vulture (*Gyps bangalensis*), Egyptian Vulture (*Neophron percnopterus*), King Vulture (*Sarcogyps calvus*).

Vultures Conservation in Tikamgarh: Identification

Birds of prey or Raptors are birds that hunt for food primarily on the wing, using their keen senses, especially vision.

Raptors have:

1. Very good **eye sight** for finding food,
2. **Strong feet** for holding food, and
3. Strong **curved beak** for tearing flesh.

Vultures are carrion-eating raptors of two distinct biological families, each occurring in only the Eastern Hemisphere (Accipitridae) or the Western (Cathartidae).

Members of both groups have heads either partly or fully devoid of feathers.

Vulture Species in Tikamgargh

The information is based on the survey work of last 6 years (2006-2012).

Tikamgarh has 5 vulture species (includes residential and migratory spps):

1. *Gyps indicus* (Long-billed vulture)
2. *Gyps bengalensis* (White-rumped vulture/Oriental White-backed vulture)
3. *Sarcogyps calvus* (King vulture/Red-headed Vulture)
4. *Neophron percnopterus* (Egyptian vulture)
5. *Gyps fulvus* (Eurasian Griffon vulture)

Neophron percnopterus (Egyptian Vulture)

Size: 65-75 cm

Colour: dirty-white raptor

Nest: on cliffs, trees, buildings

Distribution: Fairly common breeding resident throughout region. Also occurs in S Europe, Africa and W Central and E Asia.

Habits: Probably now the commonest vulture in the region. Soars over wide areas in search of food. Often perches on ruins, and high roofs.

Sexes alike.

Identifying features: Wedge-shaped tail. White plumage often stained rusty. Small, shaggy head with yellow face and thin yellow or grey bill. Immature blackish brown with grey face.

Gyps indicus (Indian Vulture or long-billed)

Size: 80-95 cm.

Colour: pale brown raptor with dark flight feathers.

Nest: cliffs, monuments, trees in exceptional cases.

Distribution: Globally threatened. Most frequent in large wild life sanctuaries.

Habits: Spends much time soaring, covering huge areas. Waddles on ground and sits hunched.

Identifying features: white thighs and ruff and bare dark brown head and neck with scattered white fluff. Broad wings, small head, long neck and short tail.

Sexes Alike.

Immature even paler with pale head and neck and grey bill.

Gyps bengalensis (Oriental White-Backed Vulture/White-Rumped Vulture)

Size: 75-85 cm

Colour: dark brown

Nest: Big trees

Distribution: Globally threatened. Most frequent in Northern wildlife sanctuaries and now very scarce in towns.

Habits: Feeds on dead livestock and predator kills. Spends much time soaring, covering huge areas. Waddles on ground and sits hunched. Very sociable where it can be.

Identifying features: broad-winged raptor with large white back patch. Adult blackish-brown with whitish ruff and underwing coverts and brownish-grey, naked head and neck. Broad wings, small head and short tail. Bill blackish. Sexes similar. Immature lighter streaked brown with paler neck and no underwing contrast.

Gyps fulvus (Eurasian Griffon)

Size: 95-105 cm

Colour: huge, rufous-brown, white neck ruff on adults.

Nest: Nests in small groups on cliff edges.

Distribution: Common breeding resident of western Pakistan and northern India and Nepal. Also occurs in S Europe, Africa and W Central and E Asia.

Habits: Inhabits mountains and semi-desert, wandering into nearby dry plains in winter. Usually in pairs or small groups, often with other vultures. Spends much time soaring looking for carcasses.

Identifying features: Darker and smaller than Himalayan. White head and neck. Bill yellow in adult, grey in immature. In flight shows dark, rather streaked, body, dark flight feathers and tail and pale underwing coverts with some white lines. Sexes similar.

Sarogyps calvus (Red-headed Vulture or King Vulture)

Size: 85 cm

Colour: blackish raptor with red head and neck

Nest: Nests high in a lone tree, often near a village.

Distribution: Scarce but widespread breeding resident throughout the lowlands.

Habits: Usually singly or in pairs. Inhabits cultivation, semi-desert and open woodland.

Identifying features: white breast and thighs. Adults have bare, red head and neck and reddish feet. Eyes and cere yellow, sharply hooked bill, black. Sexes alike. Immature browner with white down on head. Immature have white belly and pink head.

Recommendation

Conservation of Goushalas and emphasis on census of livestock Tikamgarh and Jhansi.

Assessment of number of slaughter houses in and around Tikamgarh.

Prohibition on cutting of huge trees like Sembhal tree, Arjun tree, Neem, Bargad and Haldu tree etc. Emphasis on plantation of the mentioned large trees.

Prohibition on picking and collection of small twigs from the conserved and protected areas.

Prohibition on use of loud speakers and D. J. in nearby area of Orchha monuments (especially in Orchha resorts and tourism hotels).

Prohibition on use of crackers and fireworks in and around Orchha monuments on festivities.

Prohibition on entries of tourists inside the monuments especially on the upper floors.

Putting of warning boards in and around Chhatris regarding prohibition on pelting stones on vultures and on their nest.

Establishment of vulture restaurant *i.e.* dumping site for dead carcasses in Maror *i.e.* the areas under the supervision of the forest department as well as far away from the city. For this purpose local people can be promoted and awared as well as a Veterinary Doctor can be directed for testing of carcasses.

Training programme for veterinary doctor as well as forest guards for the rescue operations of injured and dehydrated birds.

Establishment of one room rehabilitation centre near the breeding site under the supervision of the forest department.

An exhaustive mass awareness programme to sensitize the local people such as students, teachers, businessman, agricultural, people related to temples and others also.

Total prohibition on mending of the monuments during breeding season *i.e.* September- June by coordinating with Archeological department.

Total prohibition on permission for film shooting inside the premises of the Orchha monuments during breeding season *i.e.* September- June by coordinating with archeological department and office of the District Magistrate.

Proper coordination with the mining department to postponed the mining activities during the breeding season.

To develop a vulture safe zone in Tikmgarh *i.e.* the area about 200 Km in and around the breeding colony of vulture should be free from harmful anthropogenic activities.

Conservation of all feeding and roosting site of vultures.

A regular census (at the gap of 6 months *i.e.* in month of December and May) as well as monitoring of roosting and breeding site can be done by the forest guards.

Any research on vultures in the Tikamgarh area should be allowed only after verifying the proper permission from the PCCF Wild life Bhopal, and a copy of synopsis should be submitted. After the completion of work, a report of the research should be submitted to the PCCF Wildlife.

Threat due to diclofenac poisoning

Diclofenac poisoning has caused the vulture population in India and Pakistan to decline by up to 95% in the past decade, and two or three of the species of vulture in South Asia are nearing extinction. This has been caused by the practice of medicating working farm animals with diclofenac, which is a non-steroidal anti-inflammatory drug (NSAID) with anti-inflammatory and pain killing actions. Diclofenac administration keeps animals that are ill or in pain working on the land for longer, but, if the ill animals die, their carcasses contain diclofenac. Farmers leave the dead animals out in the open, relying on vultures to tidy up. Diclofenac present in carcass flesh is eaten by vultures, which are sensitive to diclofenac, and they suffer kidney failure, visceral gout, and death as a result of diclofenac poisoning.

The decline in vultures has led to hygiene problems in India as carcasses of dead animals now tend to rot, or be eaten by rats or wild dogs, rather than be tidied up by vultures. Rabies among these other scavengers is a major health threat. India has one of the world's highest incidences of rabies.

The decline in vultures causes particular problems for certain communities, such as the Parsi, who practice sky burials, where the human dead are put on the top of a Tower of Silence and are eaten by vultures, leaving only dry bones.

Meloxicam (another NSAID) has been found to be harmless to vultures and should prove an acceptable alternative to diclofenac. The Government of India banned diclofenac, but over a year later, in 2007, it continued to be sold and is still a problem in other parts of the world.

Chapter 13

Impact of Environmental Degradation on Fisheries

Ashwani Kumar Dubey

Research and Development Unit,
Godavari Academy of Science and Technology,
Under the Wing of Environment and Social Welfare Society,
Chhatarpur – 471 001

ABSTRACT

Environmental degradation is the deterioration of the environment through depletion of resources such as air, water and soil; the destruction of aquatic ecosystem. As humans dump waste products, use chemicals, and over fish in the aquatic water bodies, areas of beauty such as beautiful aquatic plants and animals are damaged. At times the destruction is so great that is cannot be reversed. We are killing our planet and the consequences are tremendous. Now a day's freshwater resources at high risk. Vast areas of fisheries have been lost. Aquatic plant and organism continue to suffer degradation, and the total fish available for catching is declining. Not only is the degradation causing aquatic resources to be lost, but this issue holds large economic problems. When there are not enough fish to catch, fishermen are without income to support themselves and their families. In this area various livelihoods such as fishing, vegetable production and traditional betel leaf farming are facing one of the worst crises ever. Therefore, it is a prime time to discuss and solve the water crises for fisheries as well as economic problems.

Keywords: *Fisheries, Water resources, Environmental degradation, Climate change.*

Introduction

Environmental degradation is the deterioration of the environment through depletion of resources such as air, water and soil the destruction of ecosystem. It is defined as any change to the environment perceived to be undesirable (Johnson *et al.*, 1997). Climate change affects the Earth's water supply in a large number of ways. Our land, water and soil are compromised when people exhaust resources or release harmful chemicals into the air. Deforestation, wasting resources, and pollution all add to the demise of an environmentally-sound and safe planet. One major component of environmental degradation is the depletion of the resource of fresh water on Earth. Approximately only 2.5 per cent of all of the water on Earth is fresh water, with the rest being salt water. 69 per cent of the fresh water is frozen on Antarctica and Greenland, so only 30 per cent of the 2.5 per cent of fresh water is available for consumption. Fresh water is an exceptionally important resource, since life on Earth is ultimately dependent on it. Water transports nutrients and chemicals within all forms of life, sustain both plants and animals, and mould the surface of the Earth with transportation and deposition of materials (Young *et al.*, 2004).

The available fresh water being affected by climate is also being stretched across an ever-increasing global population. It is estimated that almost a quarter of the global population is living in an area that is using more than 20 per cent of their renewable water supply; water use will rise with population while the water is also being aggravated by decreases in stream flow and ground water caused by climate change. Even though some areas may see an increase in freshwater supply from an uneven distribution of precipitation increase, an increased use of water supply is expected. The issue of the depletion of fresh water can be met by increased efforts in water management (Young *et al.*, 2004). While water management systems are often flexible, adaptation to new hydrologic conditions may be very costly Ragab and Christel (2002). Preventative approaches are necessary to avoid high costs of inefficiency and the need for rehabilitation of water supplies (Young *et al.*, 2004) and innovations to decrease overall demand may be important in planning water sustainability (Tilman 2011).

Water supply systems, as they exist now, were based on the assumptions of the current climate, and built to accommodate existing river flows and flood frequencies. Reservoirs are operated based on past hydrologic records, and irrigation systems on historical temperature, water availability, and crop water requirements; these may not be a reliable guide to the future. Re-examining engineering designs, operations, optimizations, and planning, as well as re-evaluating legal, technical, and economic approaches to manage water resources are very important for the future of water management in response to water degradation. Another approach is water privatization; despite its economic and cultural effects, service quality and overall quality of the water can be more easily controlled and distributed. Rationality and sustainability is appropriate, and requires limits to overexploitation and pollution, and efforts in conservation (Young *et al.*, 2004).

Observation and View

Water is the essential part of our life as well as of the environment. When factories produce harmful chemicals and toxic waste into bodies of water, humans suffer. Pesticides and fertilizers can also get into a region's water system and pollute it; drinking water is contaminated. Some residing in third-world countries are highly affected by the degradation of our planet and these unhealthy practices cause *viz.* Illnesses, Death in children and Death in adults. In the Chhatarpur district here freshwater resources at high risk. Vast areas of water bodies have been lost. Adverse effect of environment, climatic changes, increasing water temperature (Parihar and Dubey,1995), declining water level (Dubey *et al.*, 2011), tremendous use of pesticide and xenobiotic compound (Dubey, 1995), Heavy metals (Yusuf *et al.*, 2014) routines dumping of city garbage and garlanding in the aquatic bodies affected the fisheries productivity hence decreasing the number of aquatic organism specially fish species in the aquatic ecosystem. Aquatic plant and organism continue to suffer degradation, and the total fish available for catching is declining. Not only is the degradation causing aquatic resources to be lost, but this issue holds large economic problems. When there are not enough fish to catch, fishermen are without income to support themselves and their families.

Conclusion

In this area various livelihoods such as fishing, vegetable production and traditional betel leaf farming are facing one of the worst crises ever. Therefore, it is a prime time to solve the water crises for fisheries as well as economic problems by safe step taken for water conservation.

References

1. Johnson, D. L., Ambrose, S. H., Bassett, T. J., Bowen,M. L., Crummey, D. E., Isaacson, J.S. Johnson, D.N., Lamb,P., Saul,M. and Winter-Nelson, A. E. (1997) Meanings of environmental terms. *Journal of Environmental Quality* 26: 581–589.
2. Young, G. J., James Dooge, and John C. R. (2004). Global Water Resource Issues. Cambridge UP.
3. Raleigh, C. and Henrik, U (2007). Climate Change, Environmental Degradation, and Armed Conflict. *Political Geography*. 26. 6 674–94.
4. Ragab, R, and Christel P. (2002). Soil and Water: Climate Change and Water Resources Management in Arid and Semi-Arid Regions: Prospective Challenges for the 21st Century. *Biosystems Engineering* 81: 1, 3-34.
5. Tilman, D., Joseph F., Brian W., Carla D'A., Andrew D., Robert H., David S., William S., Danielle, S., and Deborah S. (2011). Forecasting Agriculturally Driven Global Environmental Change. Science. 292. 5515, 281-84.
6. Parihar, M. S. and Dubey, A. K. (1995). Lipid peroxidation and ascorbic acid status in respiratory organs of male and female freshwater catfish *Heteropneustes fossils* exposed to temperature increase. *Comparative Biochemistry and Physiology*.11 2C(3), 309-313.
7. Dubey, A. K., Singh, G. and Chauhan, A. (2011). Decline resource of water at Chhatarpur. M.P. National Seminar Bioresources Management, India, Sagar. 73.

8. Dubey, A. K. (1995). Responses of Antioxidants, Lipid-protein Interactions and Lipid Peroxidation in *Heteropneustes fossilis* to oxidative damage exposure. Ph. D. Thesis, Vikram University,Ujjain, India.

9. Yusuf, K., Yusuf, L. and Misra, N (2014). Effect of heavy metals (Cu, Zn) on the growth of *Catharanthus roseus*. *International Journal of Global Science Research*. 1(1), 2014. 87-95.

Chapter 14

Biotechnology and IPR

Bharat Kumar Singh and Shreaa Nath Singh

Natoinal University of Study and Research in Law, Ranchi, Jharkhand

ABSTRACT

Biotechnology has been among one of the oldest sciences that has aided the human civilization. Slowly and gradually with the advancement in the technological era, the science has undergone magnificent changes. Since the immense risk of infringement and the manner it is hindering the progress of biotechnological innovations cannot be side lined in the modern world, the IPR protection seems indispensable. Accordingly, the IPR regime of any country must be exhaustive enough so as to figure out any complication smoothly. Defining clear IPR policies is the need of the hour. In India, the patents are granted in order to secure and encourage inventions on a commercial scale however they do cover the human right aspect by checking that the patentee is not enjoying monopoly over the patented thing. The paper arguments on the inadequate protection to the biotech patents that has remained a substantial issue over the past time. The refreshment of laws seems quite mandatory in observation of the present day incompetent legal norms.

Introduction

With the progress of science and technology, human civilization is touching new horizons of development. The advancement in the field of biotechnology is influencing the progress tremendously. Biotechnology[1] is that scientific knowledge that utilizes living organisms for the commercial purpose. It is considered as one of the oldest sciences.[2] The *World Intellectual Property Organization* (WIPO) defines

1 The word bio is taken from Greek word 'bios' which means life. Technology means application of scientific knowledge for practical purposes to get desired results.

2 For example the production of wine or beer, cheese or curd, all these products involves the utilization of micro-organisms.

biotechnology as any technology using living entities, in particular animals, plants, or microorganisms, or causing change in them."[3] However the way the modern technology has developed, it can be differentiated from the age old fermentation technology.[4] Whether we talk about its applications in the Industrial area or the health care sector, whether its crop production or agriculture, biotechnology[5] is involved everywhere. The science is also applied for the environmental uses (biodegradation processes etc).

Very similar to other fields of technology, legal protection is also needed in the biotechnology sector as the processes that are involved in it are none other than human creations and human creativity when applied and practiced is in great need of being protected from the rising threats of biopiracy and exploitation of *traditional knowledge*[6]. Patent laws in most of the countries are tuned for non-biological material.[7] However biotechnological inventions[8] do need legal protection.[9]

Critically, biotechnology is divided into three of its branches. While the *'red biotechnology'* deals with drug discovery, *'green biotechnology'* deals with the agriculture. Industrial biotechnology is dealt under *'white biotechnology'*. Principally, wide research part dealt under each branch can be protected by IPR. The protection is not only in the interest of the investors but also in the interest of the public.

---

3 Graeme T. Laurie, Biotechnology and Intellectual Property: A Marriage of Inconvenience?, in CONTEMPORARY ISSUES IN LAW, MEDICINE AND ETHICS 237, 238 (Sheila A. M. McLean ed., 1996) (citing Committee of Experts on Biotechnology Inventions and Industrial Property, Second Session (Geneva, Feb. 3-7, 1986), reported in INDUSTRIAL PROPERTY, June 1986, at 251, 256.

4 Some of the procedures involved in modern technology include the DNA techniques, the cloning methods and the well known nano technology. The modern biotechnology is aged to be 50 years old only.

5 The Organisation for Economic Co-operation and Development (OECD) has defined biotechnology as: 'The application of science and technology to living organisms, as well as parts, products and models thereof, to alter living or non-living materials for the production of knowledge, goods and services as available at http://www.stat.fi/tk/yr/ttbio_en.html?tulost visited on 16th September, 2005 accessed on 6th Jan, 2015

6 *The term Traditional Knowledge refers to the content or substance of knowledge resulting from Intellectual activity in a traditional context, and includes the know-how, skills, innovations, practices and learning that form part of traditional knowledge systems and knowledge embodying traditional lifestyles of indigenous and local communities, or contained in codified knowledge systems passed between generations. It is not limited to any specific technical field and may include agricultural, environmental and medicinal knowledge, and knowledge associated with genetic resources-Intergovernmental Committee on Intellectual Property and Genetic Resources, Traditional Knowledge and Folklore*, WIPO, available at http://www.wipo.int/edocs/mdocs/sct/en/wipo_grtkf_ic_17/wipo_grtkf_ic_17_inf_9.pdf accessed on 6th Jan, 2015

7 H.S. Chawla, Patenting of Biological Material and Biotechnology, Journal of Intellectual Property Rights, Vol 10, January 2005, pp 44-51 available at nopr.niscair.res.in/bitstream/123456789/.../1/JIPR%2010(1)%2044-51.pdf accessed on 6th Jan, 2015

8 Patent laws under intellectual property rights make for the protection of biotechnological inventions.

9 http://www.palgrave-journals.com/jcb/journal/v14/n1/abs/3050079a.html, accessed on 5th Jan, 2015

The next major question arises that how IPR deals with the subject matter of biotechnology? In India, there exist no separate laws on biotechnology however the IPR laws of the country comprise of a large web of rules and regulations on patent protection, trademarks, registered designs, trade secrets, plant breeders' rights, domain names etc. These separate legislations make a possible connection to the science of biotechnology and hence work for its protection.

The Benefits of Biotechnology cannot be Sidelined

Ranging from *biomedical technologies*[10] to the *agricultural sector*[11], from *marine biotechnology*[12] to the *environmental biotechnology*[13], the wide benefits cannot be sidelined in the modern world as it coddles the need of the society. Hence in order to secure this form of science to its fullest, IPR protection is indispensable. However what raises the controversy is the dearth of entrenched practice. Again the Indian tradition does not allow fixing monetary value to anything that is intangible and engages no market value.[14] Hence defining clear IPR policies is not only a need but also a necessity.

Table 14.1: Link between Biotechnology and IPR[15]

PATENTS (Patents Act, 1970)	Basically work for the protection of medical devices, methods for preparation or use of enzymes, microorganisms, viruses, antibodies, vaccines, plants, seeds etc.
TRADEMARKS (Trademarks act, 1999)	Work for the protection of words or names, computer icons, graphical designs, multimedia elements or use of the above.
REGISTERED DESIGNS (Designs Act, 2000)	Medical devices, biochemical, biophysical or bio-electrochemical apparatus are covered under this.
TRADE SECRETS (Uniform Trades Secrets Act, 1970)	Involve the protection of Laboratory notebooks, customer information, documented internal processes etc.
PLANT BREEDERS' OR PLANT VARIETY RIGHTS (Protection of Plant Varieties and Farmers' Rights Act, 2001)	Involve the protection of Plant varieties, propagating and harvesting material from plant varieties.
DOMAIN NAMES	Involve the protection of web addresses.

Legal Regulation of Biotechnology

The ***Indian Patents Act, 1970*** is based on the philosophy that patents must be granted to secure and encourage inventions on a commercial scale. However the law also keeps a check that the patentee is not enjoying monopoly over the patented

10 Recombinant drugs, recombinant diagnostic kits and vaccines.

11 Producing GM crops- more pest resistant and high nutritional qualities.

12 Fish farming.

13 Treatment of air pollution.

14 http://www.ciesin.org/docs/008-265/ch31-bio.html, accessed on 5th Jan, 2015.

15 Referred from "Biotechnology Intellectual Property management manual", Spruson & Ferguson, 2008.

thing. The Indian patents act has been amended several times to make it competent according to the terms of TRIPS[16] agreement. India being a signatory to WTO's TRIPS agreement was put under the contractual obligation to amend its Patents Act in compliance with the provisions of TRIPS. Accordingly the Indian Patent Act, 1970 was amended in 1999, 2002 and 2005 to meet the requirements of TRIPS. The discussion with respect to ***EUROPE*** has been limited to the European Patent Convention (EPC). As per the European Patent Convention, any invention is patentable unless it falls within the list of excluded inventions.[17] According to Article 52 of EPC, any invention irrespective of the technology to which it belongs can be considered as patentable subject matter so long as it is new, inventive and has an industrial applicability and does not fall within the list of excluded inventions provided in Article 53 of the EPC.[18] To be patentable subject matter in **USA**, an invention should be a process, machine, manufacture or composition of matter or any improvement thereof.[19] The biotechnological inventions have been lowered to certain extent in the U.S.A. Courts from time to time have held that purified 'gene sequences' even if they match what has been created by the nature fit the definition of novelty.[20]

Major Provisions in Relation to Biotechnology and some of the Persistent Issues

a. *Article 27* (*1*) of the agreement provides that "patents shall be available in all fields of technology, without discrimination, provided that they are new, involve an inventive step and are capable of industrial application".

b. Again *Article 27*(2) enables a Member to exclude from patentability inventions whose commercial exploitation may be contrary to public order or morality.[21]

c. Further, Article 27(3) allows Members to exclude from patentability certain subject matter, such as plants and animals.

Criteria for Patenting

The general patentability requirements include novelty, industrial applicability (utility), non-obviousness and disclosure of the invention. However there has been an intense debate on this subject matter. (1) The criteria of industrial application is challenging to prove in the case of inventions in biotechnology. Again (2) benchmarks of novelty and non-obviousness are very difficult to be established in the case of

---

16 The *Agreement on Trade Related Aspects of Intellectual Property Rights* introduced intellectual property law into the international trading system for the first time and stands as the most comprehensive international agreement on intellectual property till date.

17 Article 52, European Patent Convention

18 Article 52, European Patent Convention 1973 as amended in 2000

19 35 USC Section 101 (2005)

20 It has been held that isolation and purification of a naturally existing gene sequence lends novelty to the sequence; as taken from Dr. Kalyan C. Kankanala, *Patenting Biotechnology Inventions* available at www.slideshare.net/Brainleague/introduction-to-patent-law accessed on 6th Jan, 2015

living organisms. Now in various countries, it is equally considerable that if the invention does not exists in any of the prior art, then it is suited for the protection procedure. The "prior art" includes anything which, under the statute, can destroy novelty (a result known as 'anticipation') or render the invention "obvious".[22] To determine novelty, it is necessary to first identify the prior art and then individually compare each prior art reference with the claims. Again the disclosure criteria is met as has been enshrined in the Budapest Treaty merely by depositing microorganisms in any of the internationally recognized depository.

Inadequate Protection to Biotech Patents

A large no. of industries are extracting the benefits of sound human resource, vast bio-diversity and large domestic market in the Indian platform. These enormous characteristics also result into the low cost research as compared to other countries and that makes it the primary reason what fascinates the international companies to start their venture at the Indian markets. However this particular actuality raises questions to one peculiar question that *"Is Indian Patents law framework really compatible according to the needs of the nation in relation to Biotech Patents?"* This concern is high lightened because though India remains the member to various international treaties and agreements still momentous enhancement in the areas of operation and enforcement of patent laws is the need of the hour.[23]

Invention vs Discovery

While it comes to the issue of securing Biotech patents, the seperatiblity of a *'Discovery[24]'* from an *'Invention'* stands as a major form of debate. Whether finding new 'life forms' be counted as a discovery or an invention can be gathered a hint by making the differentiation between the two. A substance freely occurring in nature, if merely found or discovered, is not patentable.[25] However, if the substance found in nature has first been isolated from its surroundings and a process for obtaining it, is developed, that process is considered invention and hence patentable. But the field of biotechnological patents are not only restricted to the 'discovery' part. The inventions regarding production of replicas, parts of organs, tissues or cells, other biotechnological innovations etc do count as *'biotechnological inventions'* and such are smoothly granted protection under the head of law. Hence it is a critically dealt dispute as to what constitutes an invention in contrast to a discovery.

---

21 The same ideology has been maintained in European Patent Convention, Article 53 (a). The article reads denies patentability to "inventions, the publication or exploitation of which would be contrary to "ordre public" or morality.

22 Introduction to Novelty, Nonobviousness, and Double Patenting- Chapter 4, Biotechnology and the Law, Volume 1.

23 http://russianlics.sstu.ru/globelics.nsf/0/18CC45447D998BFFC32572CF003EBCC8/$File/Globehelics%20-%20P.Arora%20(India).doc, accessed on 6th Jan, 2015

24 Discovery means merely making available what already exists in nature.

25 http://ijme.in/index.php/ijme/article/view/1168/2643

What is Patentable and What is Not

The US Supreme Court in *Diamond v. Chakrabarty*, a landmark biotech case, held that everything under the sun made by man is patentable.[26] However, as per the arguments raised by several Civil Society Organizations (CSO) and Non-Governmental Organizations (NGO), what has been created by the God should be left for the application by the entire mankind and hence cannot be appropriated by any particular individual, organisation or group.[27] *Iver P. Cooper* explains this in terms of the "product of nature" doctrine. This doctrine states that a thing occurring in nature which is substantially unaltered is not a manufacture and is therefore not patentable.[28] The term 'manufacture' refers to something made by the hand of man, and not by nature alone.[29] These contenders do not see much advancement and innovation being done in the field of biotechnology and hence secure the idea that life form innovations must not be patented. They are also the proponent of the belief that all human beings are equal before each other and carry equal rights. No one should be granted higher rights than the other.

The Requirement of Disposition

No patent protection shall be availed unless the invention has been made available in the public. Hence before any protection can be claimed, the work has to be made public so as to check any infringement. Such disposition requirement has been drafted in the *Budapest Treaty*[30] of 1977 which requires a single deposit made at any of the depositories will suffice. However such public availability surely does includes the competitors and hence the chances of emergence of similar work cannot be estopped. This can be suitably described as a *'pocket factory handed over to the imitator on a silver plate'*.[31]

Biopiracy: A Threat to Biotechnological Advancement

The term 'biopiracy' normally refers to the unauthorized extraction of biological resources and associated traditional knowledge from developing countries without compensation of the fake inventions that have been generated in application to that knowledge. Intimidation of the work of the poor farmers and indigenous people of the developing countries by the private entities or corporations belonging to the

---

26 Diamond v. Chakrabarty, 100 S.Ct. 2204 (1980); It also stated that living organisms were not excluded from the scope of patentable subject matter in USA

27 http://www.iprhelpdesk.eu/sites/default/files/newsdocuments/IP_in_Biotechnology.pdf, accessed on 5th Jan, 2015

28 Iver P. Cooper, Biotechnology and the Law- Volume 1, 2014

29 Oxford English Dictionary: "Manufacture 1.a The action or process of making by hand, the process of making articles by mechanical power, the most perfect manufacture is that which dispenses entirely with manual labour."

30 Budapest Treaty on the International Recognition of the Deposit of Microorganisms for the Purposes of Patent Procedure on April 28, 1977 and into force on August 9, 1980.

31 http://www.voiceofresearch.org/Doc/Jun-2014/Jun-2014_9.pdf, accessed on 5th Jan, 2015

developed countries is leading to exploitation at large scale.[32] Such practice leads to the transfer of the patent right that basically decides the exclusive right to produce and sell the patent product. Hence in effort that the endeavor of these communities does not go in vein, the IPR system of granting patents must be made stricter.[33] These corporations at the first instance start making huge revenues by the utilization of the patents, they further come out as a major source of threat by selling the products at a much higher price.[34] The major defeat of the protection laws can be understood in way that the local communities stand unrewarded. The present IPR system is proving supportive only to the private multi-national corporations who are aware of such protection schemes and legal aids hence at the first place they register the patent in their name.[35] While there is a great need to redefine the guidelines and policies for the implementation of the IPR, wide scale awareness also needs to be ensured. Such efforts can help prevent bio-piracy.

Conclusion

India is a country where legal norms have developed with respect to ethics and morality. It is into the Indian tradition and culture to worship plants and animals. Moreover all that is created by the God is said to be the property of all hence such things are not appreciated for the purpose of patent protection. The Indian patents act does adhere to the moral ethics. It clearly defines the law in terms that "*any invention, the exploitation of which is against public order and morality and that may cause serious prejudice to the health of human beings, animals or to the environments cannot be patented.*" There certainly resides a great connection between the science of biotechnology and India as the country is preferred over others for its abundance of resources, cheap labour and lesser regulation of laws.[36] In recent years, the Biotech Industry has achieved its pace and is the fastest growing. There has also been a hike in the investments in the present sector. Considering the tremendous growth of the industry, the need for proper and effective laws seems obvious. India is considered a rich country of Traditional Knowledge[37] and this factor makes it the priority of the

---

32 Such is a common experience that is faced by the Indigenous people who always face difficulty in protecting their knowledge against the piracies occurring.

33 http://www.fbae.org/2009/FBAE/website/our-position-ipr.html, accessed on 6th Jan, 2015

34 Monopolizing the profit drawn out of the application of traditional knowledge in the sector of pharmaceutical firms is a common trend.

35 It should be noted that IPR's are primarily based on the concepts and standards that have been developed by few western countries. These concepts and standards, it has been realized, not necessarily fulfill the IP requirements or standards of majority of the developing countries. Hence as a result the codified or non codified Traditional Knowledge has also become an open treasure for misappropriation.

36 http://www.mondaq.com/india/x/26049/Healthcare/Biotechnology+Laws+in+India accessed on 6th Jan, 2015

37 The linkage between biodiversity, traditional knowledge and IPR is the product of the exploitation done by big pharmaceutical companies and the loss of various indigenous plant varieties from the forest. Reading *CBD Article 8(1)(j)* illustrates that the International instruments are having some focus towards commercializing the TK or sharing the economic benefits hence derived from it. The growing commercial importance of Biodiversity increases the necessity of framing Indigenous people centric mechanism to protect from exploitation.

developing countries. However, all such entries must be well aware of the legal norms that are persistent in the country in regards to biotechnology. While the number of biotech patent applications day by day is increasing, such have to be interpreted under the new definition which must comprise of a mental rethought over the International policies and mechanisms that are running parallel in other countries. The refreshment of laws seems quite mandatory in observation of the present day incompetent legal norms. Today, though the field is under issues, the biotechnological inventions have contributed towards the progress of the society in a large efficient manner and such serving the society shall continue in the near future. This science has the potential to furnish the needs of the society.

अध्याय 15

जलवायु परिवर्तन एवं ग्लोबल वार्मिंग

मयूर शिखा अग्रवाल

शा. राज्य स्तरीय स्नातकोत्तर विधि महाविद्यालय
बरकतउल्लाह वि.वि. भोपाल, मध्य प्रदेश

व्यवसायिक विकास जनसंख्या वृध्दि, जंगलो व खेतों का स्थान सीमेंन्ट की इमारतों द्वारा अधिग्रहित होना पर्यावरण के संतुलन को क्षति पँहुचा रहा है। जिसके कारण जलवायु में परिवर्तन तथा तापमान वृध्दि से विभिन्न स्वास्थ समस्यायें उत्पन्न हो रही है।

औद्योगिक क्रांति में उर्जा की जरूरतो की पूर्ति के लिये तेज रफतार से जीवाश्मों को जलाया जा रहा है। "इण्टर गर्वमेन्टल पैनल ऑफ क्लाइमेंन्ट चेंज" के अनुसार तापमान में बढोतरी के परिणाम स्वरूप सन् 2030 तक विश्व में समुद्र के पानी का स्तर 20 से.मी. तक और अगली सदी के अंत तक 65 से.मी. तक बढ जायेगा । परिणाम स्वरूप 300 प्रशांत प्रवालद्वीप पूरी तरह से मिट्टी मे मिल सकते है"। जलवायु से प्रभावित होते है फसल की पैदावार फूलो का खिलना सम्पूर्ण वनस्पति अनूकूलता में ही पनपती है। हमारे वर्तमान समय में मानव प्रजाति लगभग आठ अरब मीट्रिक टन कार्बन डाइऑक्साइड प्रत्येक वर्ष वायुमंडल में प्रवेश करवा रही है। कार्बन डाइआक्साइड की अधिकता से तापमान में अत्यधिक वृध्दि हो रही है। क्योकि कार्बन डाइआक्साइड की मोटी परत परावर्तित होने वाली उर्जा को वायुमण्डल से बाहर नही जाने देती है। यह ग्रीन हाउस वन जाता है। सूरज की किरणे वायुमंडल के अन्दर प्रवेश करने पश्चात उत्सर्जित होकर बाहर निकल नही पाती निष्कर्ष यह निकलता है। ग्रीन हाउस का तापमान बढ़ जाता हैं इसी को ग्लोबल वार्मिंग कहते है।

हरित ग्रह के कारण पृथ्वी का तापमान 0.8 से 2.5 तक बढ़ जाता हैं और एक अनुमान के अनुसार अगली सदी के अन्त तक पृथ्वी के तापमान में उसे 3 डिग्री सेन्टीग्रेट सें 5 डिग्री सेन्टीग्रेट की वृध्दि होने की संभावना है। ग्लोबल वार्मिंग का असर सिर्फ किसी एक देश में नही है। बल्कि पूरे विश्व पर है।

ग्लोबल वार्मिंग के दुष्प्रभाव

1. वृक्षो में समय से पहले फल आ जाना और उचित समय आते आते झड़ जाना।
2. भूमि की उत्पादन क्षमता में कमी आ जाना। जिसे दूर करने के लिये रासायनिको का सहारा लेना पड़ता है। जो स्वयं की मनुष्य के स्वास्थ पर तरह तरह से बुरा असर डालते है।
3. पशु पक्षियों के प्रजनन समय में अनियमिता।
4. पौधो की प्रकाश संशलेषण कियाओ पर प्रभाव।
5. मानसून में परिवर्तन।
6. ध्रुवो पर वर्फ पिघलना।

ग्लोबल वार्मिंग के दृष्प्रभावो के परिणाम

मौसम में आर्द्रता का समय पर न आना इससे ग्रीष्म ऋतु अधिक समय तक बनी रहती हैं। अधिक गर्मी त्वचा के कैंसर उत्पन्न करती है।

गंगा जैसी नदीयाँ जो पहाड़ की वर्फ पिघलते रहने से बनी रहती है। अधिक यमय तक तापमान बढ़े रहने से जहां एक और तटीय क्षेत्रों में बाढ़ का प्रकोप होगा तो दूसरी और सूखा दोनो ही स्थिति में मानव जीवन के लिए खतरा है।

ग्लोबल वार्मिंग का प्रभाव जानवर, मनुष्य, जीव, जन्तु, वनस्पति, वन, मानसून, पादप क्रियाओं सभी पर पड़ता है।

मनुष्य और उसके चारो ओर का जैव - अजैव का आवरण परस्पर एक दूसरे के लिये आवश्यक हैं उनकी परस्पर संतुलित अवस्था उनके अस्तित्व का आधार है।

निष्कर्ष एवं सुझाव

"पृथ्वी के पास मानव की जरूरतों को पूरा करने के लिये काफी कुछ है। लेकिन उसके लोभ को पूरा करने के लिये कुछ भी नहीं है। अगर मानव इस बात को गम्भीरता से ले तो समस्त मानवता को महाविनाश से रोका जा सकता है।" महात्मा गांधी के द्वारा किया गया उपरोक्त उद्‌वोधन आज के परिवेश में सटीक सीख है।।

निम्न सुझाव इस दिशा में सहायक हो सकते है।

1. सौर उर्जा का उपयोग कोयला एवं पेट्रोलियम के स्थान पर।
2. बायोगैस का प्रयोग।
3. ईट के भट्टो को विघुतीकरण द्वारा पकाया जाना जिसमे धुये को रोका जा सके।
4. अधिक से अधिक मात्रा में पेढ पौधे लगाया जाना चाहिए। घरेलु रूप में आँवला, तुलसी हो पत्ते वाले पौधे अनिवार्य रूप से लगाय जाए।

5. पेट्रोल वाहनों का सीमीत प्रयोग एवं लेड रहित पेट्रोल का उपयोग।
6. कारखानो की चिमनियो में फिल्टर का अनिवार्य प्रयोग एंव कारखानों के आसपास पेड़ पौधों का लगाया जाना अनिवार्य हों।
7. ग्रीन विल्डिंगस का निर्माण।

उपरोक्त के अलावा अपनी जीवनचर्या में प्रकृति के प्रति संवेदनशीलता एवं सम्मान की भावना को पुनः अपनाना होगा। विलासिता पूर्ण आदत को संतुलित करना होगा तभी इस समस्या से उबर सकेंगे।

संदर्भ ग्रंथ

1. K.C. Agrawal Environmental Pollution. Nidhi Prakashan, Bikaner (INDIA).
2. चंदोला प्रेमानन्द पर्यावरण और जीव दिल्ली (भारत) हिमाचल पुस्तक भंडार।
3. पर्यावरण संबंधी विविध अधिनियम।

Index

A

Acidic rain 21
Acidity 77, 79
Aeration 62
Agricultural land 25
Agricultural sector 11, 115
Air pollution 9
Alkalinity 65, 70, 96
All India rural credit survey report 20
Antarctica 13
APHA 86
Aspergillus sp. 61
Auto-CAD method 44
Automobiles 9

B

Bacteriological analysis 71
Bicycles 9
Biochemical oxygen demand (BOD) 77, 85
Biodiversity 4, 36
Biological oxygen demand 79
Biomedical technologies10 115
Biopiracy 118
Biotech patents 117
Biotechnology 113, 114
Bosmina 91
Brachionus 91

C

Calcium 65, 83, 86, 96, 97
Calcium hardness 69
Callosobruchus chinensis 61
Carbon dioxide 1, 2, 11, 83
Carbon dioxide concentration 25
Carbon monooxide 2
Cataract 13
Centrifugation test 71
CFL 13
CH4 1
Chemical analysis 73
Chemical oxygen demand (COD) 77, 86
Chloride 65, 70, 83, 85, 96
Chlorofluorocarbons 2
Chromium 79
Climate change 109

Colour 66, 96
Community based focused meeting 43
Community based food and water security 41
Compact fluorescent lamp 2
Completed coliform test 71
Conductivity 65, 68
Confirmed coliform test 71
Conservation 107
Cooperatives 19
Criminal activities 52

D

Daphnia 91
Data Collection 43
DDT 15
Deforestation 110
Depression 55
Dichlorodiphenyltrichloroethane 2
Disposition 118
Dissolved oxygen (DO) 79, 83, 85
Diversion 25
Diversity 83
Doha peak summit 14
Drainage 4

E

Ecofriendly management 59
Ecological conditions 29
Economy 15
Electrical conductivity 96, 97
Elnino 8
Energy conservation 17
Environment 34
Environment protection 13
Environmental biotechnology13 115
Environmental degradation 1, 109
Environmental information system (ENVIS) 35
Environmental pollution index 2
Environmental protection 19
Eri rearers cooperative society (ERCS) 22

F

Fisheries 109
Flood 4
Fluoride 65, 70
Food production 30

G

Ganges river 77
Gangotri 9
Geo-position 83
Global health 1
Global warming 4, 8, 9, 21, 25, 29
Grain moisture 61
Grain quality 59
Green biotechnology 114
Green house gases 2
Groundwater 93
Gyps bengalensis 105
Gyps fulvus 105
Gyps indicus (Long-billed vulture) 105

H

Hazardous waste 1989 14
Himalayan glaciers 9
Human health 29
Hydrogen 2
Hydrogen sulphide 2
Hydrologic conditions 110

I

ICMR 71
Immune system 13
Indian farm forestry development cooperative Ltd 22
Indian farmers fertiliser cooperative (IFFCO) 22
Indian patents act, 1970 116

Industrialization 4, 13
Infestation 61
Insomnia 56
Integrated environment development strategy 23
Intergovernmental panel on climate change (IPCC) 2
International organization for Migration 7
IPR 113, 119
Iron 70
Irrigation 41

J

Japanese consumer cooperative union (JCCU) 22
Johannesburg earth peak summit 14

K

Keratella 91
Kyoto protocol 14

L

Legal Regulation 116

M

Magnesium 65, 96, 97
Magnesium hardness 69
Marine biotechnology12 115
Mercury 2
Mesocyclops 91
Methyl bromide fumigants 64
Migration 7
Mobile phone 51
Mulberry rearers cooperative society (MRCS) 22

N

N_2O 1
Nairobi summit 14
National awareness 23
Nature conservation 33
Negative impacts 52
Neophron percnopterus 105
NGOs 33, 38
Nitrate 65, 71, 79, 86, 96, 98
Nitrite 65, 71
Nitrogen 2
Nitrous oxide 2

O

Odour 96
Oxygen 70
Ozone 2
Ozone depleting substances 2
Ozone layer depletion 13, 21
Ozone layer protection act-1987 14

P

Pacific 8
Pacific ocean 8
Patenting 116
pH 65, 68, 77, 83, 84, 96
Phosphate 79, 86
Phosphine fumigants 64
Physico-chemical analysis 93
Physico-chemical status 83
Plankton 86
Plankton diversity 83
Plodia interpunctella 61
Policy implementation 35
Population 3, 17, 21
Post facto design 44
Potability 65
Potassium 96, 98
Poverty 41
Presumptive coliform Test 71, 74
Primary grain insects 60
Problematic tree analysis 47
Psychological and physical health 55

Q

Qualitative loss 61
Quality assessment 93
Quantitative loss 61

R

Radiation 56
Rain forests 3
Reo earth summit 14
Reo+20 14
Research methodology 43
Research statistics 44
Rhyzopertha dominica 61
Royal commission on agriculture 20
Rural areas 38

S

Sanitation 62
Sarcogyps calvus 105
Secondary grain insects 60
Sitophilus oryzae 61
Sitotroga cerealella 61
Skin cancer 13
Slaughter houses 107
SMART analysis 43
SO_2 1, 11
Social responsibility 20
Socio-Economic Profile 47
Sodium 96, 98
Soil erosion 26
Stockholm summit 13
Stored grain insect pest 59
Sulphate 86
Sulphur dioxide 2
Surface water 65

T

Tasar Rearers Cooperative Society (TRCS) 22
TDS 84
Teenagers 51
Temperature 3, 25, 62, 65, 66, 77, 79, 83, 84, 111
Thermal power station 17
Thermometer 13
Total alkalinity 83, 85
Total dissolved solid (TDS) 65, 66, 83, 96, 97
Total hardness 68, 83, 85, 96
Total solid 66
Total suspended solid 66
Tribolium castaneum 61
Trogoderma granarium 61
Tsunami typhoon 8
Turbidity 65, 69, 96, 98

U

Ultra violet 2
UN convention to combat desertification (UNCCD) 34
United Nations 2
UNO 1, 13
Urbanization 13

V

Vulture 101

W

Wasting resources 110
Water quality 65, 88
Water quality criteria 71
Water quality parameters 66
Water resources 109
White biotechnology 114
WHO 71, 93
Wildlife conservation 36
World intellectual property organization (WIPO) 113

Y

Yangtze river delta (YRD) 29

www.ingramcontent.com/pod-product-compliance
Ingram Content Group UK Ltd.
Pitfield, Milton Keynes, MK11 3LW, UK
UKHW020821290926
15470UKWH00003B/1303

9 789351 308812